BIOETHICS: ASIAN PERSPECTIVES

Philosophy and Medicine

VOLUME 80

The titles published in this series are listed at the end of this volume

BIOETHICS: ASIAN PERSPECTIVES

A QUEST FOR MORAL DIVERSITY

Edited by

REN-ZONG QIU

Center for Applied Ethics & Institute of Philosophy,
Chinese Academy of Social Sciences, Beijing, China

KLUWER ACADEMIC PUBLISHERS
DORDRECHT / BOSTON / LONDON

A C.I.P. Catalogue record for this book is available from the Library of Congress.

ISBN 1-4020-1795-2

Published by Kluwer Academic Publishers,
P.O. Box 17, 3300 AA Dordrecht, The Netherlands.

Sold and distributed in North, Central and South America
by Kluwer Academic Publishers,
101 Philip Drive, Norwell, MA 02061, U.S.A.

In all other countries, sold and distributed
by Kluwer Academic Publishers, Distribution Center,
P.O. Box 322, 3300 AH Dordrecht, The Netherlands.

Printed on acid-free paper

Printed in the Netherlands.

TABLE OF CONTENTS

CHAPTER 1

REN-ZONG QIU

INTRODUCTION:
BIOETHICS AND ASIAN CULTURE – A QUEST FOR
MORAL DIVERSITY

This volume, *Bioethics: Asian Perspectives*, has been compiled and edited within the context of an intensified debate on universal ethics and global bioethics. It seems to me that after the Cold War and in the process of globalization, some have been anxious to unify not only the actions but also the beliefs and value systems in biomedical and other fields under the rubric of global bioethics or universal ethics. The desire to solve global problems or issues by coordinated efforts made by all peoples and all countries in the world is understandable. However, these coordinated efforts have to be and only can be achieved by consensus between them after patient and informative dialogue, consultation and negotiation with mutual respect and mutual understanding. The final answer cannot be deduced from an overarching universal ethics or global bioethics invented by some genius philosopher or leading figure, and any such inventor may not impose his solution on other people. Nothing could prevent this kind of imposition from leading to the kind of ethical imperialism that some bioethicists in developing countries understandably worry about.

I believe there are some values shared by different moral communities or different cultures which might constitute some common ground for resolving global issues. However, this does not mean that these values constitute an overarching universal ethics or global bioethics, because these shared values can and must be interpreted and applied in different ways in different cultural contexts. For example, Confucian or Buddhist cultures may share with Judeo-Christian cultures such values or rules as "Do not kill the innocent", "Do not steal" etc. However, for Confucians, the rule "Do not kill the innocent" does not include not killing fetuses, whereas for Buddhists it includes not killing all forms of animals. For Confucians, the rule "Do not steal" does not include stealing books: Stealing books is not stealing. Obviously not all cultures share this same interpretation of "do not steal". Although these shared values are useful in practice, their content is too poor to constitute an overarching universal ethics or global bioethics.

It turns out that these shared values or rules are only minimum denominators, as A. Campbell pointed out in his 1998 presentation "Global Bioethics - Dream or Nightmare?" at the 4th World Congress of Bioethics in Tokyo. Various communities may share moral concepts, just as Newtonian and Einsteinian physics share concepts like "mass", "space" and "time", but understand those concepts to mean different (and sometimes incompatible) things.

The contributors to this volume are all non-Westerners, but many of them have been trained in the West or are familiar with Western philosophy and/or bioethics. In the process of compiling and editing the volume I have been deeply impressed by how diverse the views of these contributors are on bioethical issues even within an Asian cultural context. I believe that the diversity or pluralism of bioethical views will promote the growth of bioethics just as the late philosopher of science, Paul Feyerabend, argued that the proliferation of scientific theories promotes the growth of knowledge.

1. THE FOUNDATION OF BIOETHICS IN ASIA

The papers in Part I show that the foundation of bioethics in Asia is quite distinct from that in the West, and that it also differs between Asian cultures and subcultures. The intellectual foundation of bioethics is entrenched in or closely related to particular cultures. In Asia there are Confucian, Taoist, Buddhist, Hindu, Islamic, Christian and many native cultures. For example, in East Asia Confucianism is still very influential, and some bioethicists are trying to develop a Confucian approach to bioethics or to integrate this approach with approaches other than Western liberalism. In her paper "Confucian and Western Notions of Human Need and Agency: Bio-medical Ethics in the Twenty-First Century," Julia Tao-Lai Po-Hwa points out that health care and biomedical ethics in the twentieth century have been grounded on two broad principles, the principle of right and the principle of need. These principles provide the ethical framework for shaping our public decision-making and social legislation, particularly in welfare choices and health care policies. However, there have been serious doubts raised in recent years about the adequacy of the language of rights, particularly in bioethical and health care matters. Feminist, communitarian and other philosophers criticize the rights language as indifferent to people's actual psychological and material capacities to exercise rights; unconcerned with responsibilities; setting claimants against each other and against larger collectives; and as being a formalistic and acontextual approach to freedom and social justice. Overemphasis on individual rights and liberty as the primary principles in health care policy and biomedical decisions can easily lead to an unlimited demand on government provisions to satisfy maximum individual preferences. This in turn can lead to the disappearance of public health as a common good in society. After she compares individualistic and relational conceptions of human need and agency, Tao argues that a relational conception or paradigm, which constitutes the basic tenet of feminist care ethics and Confucian moral philosophy, is better able to allow us to talk about doctor-patient/family relationships and to bring patients' family relations into the orbit of care and concern. She argues that an awareness of these bonds between people leads to the further recognition of our mutual responsibility for one another and to insights into the necessity of sympathy and care to provide a more adequate response to human needs in bio-medical decisions.

This approach has to focus on the concept of personhood in seeking an alternative intellectual foundation for bioethics in Asia. Edwin Hui, in "Personhood and Bioethics: A Chinese Perspective," elaborates a Confucian concept of personhood. He criticizes the "substantialist" concept of personhood in the Western tradition, in which "person" is defined as a rational individual substance, and points out the inadequacy of defining human personhood in terms of higher-brain functions or similar psychological criteria in modern bioethical discourse in the West. In contrast with this, a Confucian account of personhood emphasizes both the "substance" dimension as well as the importance of "relationality" in the constitution of personhood. The human person is a psychosomatic unity and a social/relational being, a "being" always in the process of "becoming" that is carried out in and through the social context for the purpose of fulfilling social responsibility rather than self-actualization *per se*. He analyzes the fundamental concept of *ren*, which not only refers to an inner character/spiritual condition/virtue, or an aspect of outward action/conduct, but also as a dynamic process of "person making," involving mutual incorporation between the self and the other. From this perspective it follows that different views will be taken on bioethical issues such as the beginning of life, death and dying, informed consent etc.

In his paper "Foundation of Asian Bioethics," Hyakudai Sakamoto discusses the shape of an "Asian bioethics". He argues that bioethics in the Asian region might be fundamentally different from the Western pattern in its cultural, ethnological and also philosophical basis, reflecting the present day multi-cultural post-modernism. He challenges the universality of human rights for two reasons. First, he argues that human rights have neither a theoretical background nor a practical ground in Asia. He claims that the new Asian bioethics should stand on the new philosophy concerning the relation between Nature and the human being. This is a new humanism without human-centricism, a new concept of human rights and Asian ethos.

Po-Keung Ip, in his paper "Confucian Personhood and Bioethics", examines the bioethical ramifications of the Confucian non-rights-based notion of personhood in areas including abortion, treatment of newborns and doctor-patient relationships. The Confucian notion is basically a non-rights-based notion of personhood, as well as a humanistic and collectivistic moral enterprise. Since the family sits at its core, it may aptly be deemed familial collectivism. In traditional China, central to the set of Confucian norms is filial piety. As a cardinal principle, it dictates the desirable and proper human relationships within the family and beyond. As a result of the paramount importance given to filial piety, it dominates and defines all other salient human relationships. To the extent that filial piety centrally endorses hierarchical human relationships, it runs contrary to the value of equality of persons that sits at the heart of the rights-based concept of a person. To the extent that Confucian familial collectivism is inhibitory to rights and equality, it is conceptually incompatible with the concept of a rights-based personhood. Ip also explores the implications of the concept of personhood for bioethical issues such as abortion, euthanasia, organ transplant, human experimentation, and artificial human procreation. In his conclusion he points out that the chief flaw of the Confucian non-rights-based personhood is that it fails to take due consideration of the rationality of the rights arguments as well as the rights of the agents involved in decision-making.

In "Rights or Virtues? Towards a Reconstructionist Confucian Bioethics", Ruiping Fan addresses the arguments offered by Tao, Hui, Sakamoto and Ip. From Fan's view, the two sets of essays in this section, Tao's, Hui's, and Sakamoto's on the one hand, and Ip's on the other, epitomize two different approaches to the nature of Asian bioethical explorations for the foundation of an Asian bioethics. In the approach of Tao, Hui and Sakamoto, the foundation must be established based on Asian cultures, religions, and moralities, especially Confucianism, thus providing a bioethical account more adequate in the Asian context than the account offered by modern Western individualist morality. In Ip's approach, however, a rights-oriented bioethics is a must to shape the core of an Asian bioethics, because, as he sees it, Asian moralities in general and Confucianism in particular fail to take individual rights seriously and are thereby fundamentally defective. Fan argues that a strict rights-based account of bioethics is problematic, and suggests that virtue-based Confucianism has many strengths to recommend it as a possible basis of bioethics. In his 'Reconstructionist' bioethics, he points out the very different moral perspectives underlying Confucianism and the liberal morality of the West, and argues that attempts to reduce Confucian ideals to familiar Western concepts is not possible without a loss of meaning. According to Fan, "Reconstructionst Confucianism holds that it provides a more ample account of human flourishing and morality than that offered by other accounts, individualist liberal accounts included" (p. 72).

2. BIOETHICS IN ASIAN CULTURE: GLOBAL OR LOCAL?

Part II of this volume focuses on the impact of culture on the way in which bioethical issues are formulated and addressed, and in perusing this section, readers cannot help asking the question: is bioethics global or local? It seems to me that in spite of there being shared values between different moral communities and different cultures, bioethics as well as ethics in general seems to be grounded locally. However, the contributors of papers in this part of the volume disagree on this topic: The diversity of bioethical views or the debate between them is natural in this rapidly developing and changing continent which is undergoing a paradigm-shift.

Angeles T. Alora, in her paper "Philippine Culture and Bioethics," clearly describes the impact of culture on bioethics in the Philippines and the interaction between culture and bioethics. She points out that what is bioethically correct may not be culturally acceptable and similarly what is culturally normative may not be bioethically correct. We must consider this interplay and determine what is both culturally and bioethically acceptable. The characteristics of Filipino culture include society or family orientation, a search for harmony, person orientation, and multidimensional health care. For Filipinos the family is considered "the highest value in Filipino culture" and the "core of all social, cultural and economic activity", and it provides financial and psychological support, emotional security, and a feeling of belonging—an environment where a Filipino can be himself. The Filipino views his person holistically, and rarely isolates actions from person, parts (whether these be organ/cell/molecule/gene) from whole, mental from physical, disease from diseased. All these affect how such bioethical issues as autonomy, futile treatment and care for the terminally ill, health professional relationships, health policy etc. are addressed. She claims to develop from the different cultures "a basic fundamental moral attitude, a fundamental consensus of binding values, standards and personal moral attitudes, an irrevocable ethics, a universal ethic." However, when in a particular situation bioethical principles come in conflict with the cultural givens, we should learn to use prudence and patience to discern what is truly ethically correct and culturally acceptable in the Philippines.

V. Manickavel examines the encounters of bioethics with different cultures in different countries from another perspective in his paper "Living in Separate and Unequal Worlds: A Study in the Application of Bioethics". He argues that bioethics is like a train that doesn't stop at stations that may be too small or where people don't want to get on. He takes an example of American physicians using Quinacrine for sterilization of Indian women (with local government approval), despite the fact that neither the United States nor other Northern countries had approved this particular drug study for this purpose in their own countries. This example illustrates the insensitivity of the host Southern country and the double standard adopted by the Northern sponsoring countries. In conclusion he argues that ethics is not a luxury that only rich people or nations can afford. Instead, the ethical principles based on transcendent values always help humanity in providing justice to individuals, families, communities or nations.

Another interesting aspect is to look at the impact of Asian culture upon the relationship between physician and patient. In many cases this relationship is a triple relationship between physician, patient and her/his family. Because bioethics in Asia is mainly beneficence-oriented, the principles of autonomy, informed consent, truth-telling, etc., cannot be implemented in a Western way, and have to accommodate themselves to an Asian context. Un-Jong Pak, in her paper, "Medical Ethics and Communicative Ethics" addresses the limitations in applying the principle-based argumentation in the era of modern science and technology, and argues for a communicative ethics that is based on feminist care ethics integrating Oriental morality. She argues that the application of modern biomedical technology brings about possible harms to human beings, and that there is a need to reconstruct the bioethics that is based on mutual respect and co-responsibility between equal moral subjects. Modern biomedical science has changed the context of medical treatment, so that in the decision making process, the patient relies more on communication with family or others than with the physician. Eventually, most decisions are made by the patient her/himself with the support of other family members. The patient must seriously consider the non-medical factors such as the extent of financial burden on him/herself and others, quality of life, or family impact rather than medical concerns such as the effectiveness of medical care, risk of operation, etc. What matters in bioethical issues is to capture the dilemma in the situation and to suggest a solution. There lies the reason why we can not establish a satisfactory answer merely by appealing to theory or general principles in resolving bioethical issues. In this regard, the development of the new communicative model has emerged as one of the important subjects of bioethics. She argues that communication based upon mutual understanding and agreement is the essence of ethics, and the patient's right to autonomy or informed consent is deemed to be secondary. Pak claims to reconstruct a new medical ethics through the reinterpretation of traditional Asian values and the perception of the community. Her endeavor as a Korean feminist philosopher to develop a "Buddhist model of the caring ethics" and such traditional Oriental ideas as the "relational ontology" is an example of such a reconstruction.

Now let us turn to other parts of Asia. Leonardo D. De Castro & Allen Andrew A. Alvarez in their paper "*Sakit* and *Karamdamam*: Towards Authenticity in Filipino Concepts of Disease and Illness" highlight the value of authenticity in responding to disease and illness. For them authenticity must be added to utilitarian and pragmatic considerations as a criterion of validity of healing responses to disease and illness, and must be achieved by ensuring the alignment of treatments and remedies with the social and cultural dimensions of disease and illness concepts. The point is that diseases and illnesses are not merely biophysical phenomena, but instead form part of a matrix of values, traditions and beliefs that define a cultural identity. They begin by analyzing the nuances in meaning of two words, "*sakit*" and "*karamdamam*," which have noteworthy implications for the understanding of concepts of health and illness in the context of Filipino society and culture. Various theories of illness causation remind us of a need to situate the search for health remedies in the context of cultures, traditions and ideologies. In conclusion, they argue for cultural integrity and authenticity in medicine. From their arguments it may follow that there is no essential bioethics or essential medicine, as pointed out by Arthur Kleinman.

3. LIFE AND DEATH, EUTHANASIA AND END-OF-LIFE CARE

Life and death are major concerns of all peoples, and are also major issues in bioethics. Perhaps the most voluminous literature in bioethics is devoted to these two issues. And the inquiry into life and death issues highlights the cultural influence on bioethics. The advance of new genetics, cloning, reproductive technology, life-sustaining and other relevant technologies enable human beings to "play God", i.e., take over the arrangement of life, aging, disease and death from Nature or God. That makes the life and death issues more complicated. Phee Seng Kang, in his paper, "Cloning Humans? Some Moral Considerations" discusses moral issues involved in the means as well as the ends of human cloning, and explores the relationship between genetic equivalence and identity equivalence, the implication of inheriting another person's genes, the effect of asexual reproduction on family structure and relationships, the risks involved in human cloning, and the question of cloning and rights. The extreme individualism and liberalism underlying the acceptance of a woman's right to use her dead husband's frozen sperm without his consent and become pregnant more than three years after his death are not ideas to which an Oriental society will easily subscribe. He suggests that the most urgent task is not to perfect the technique for human cloning so that infertile couples can have another option to procreate, much less to establish a possible scenario that would permit human cloning. He criticizes the U.S. National Bioethics Advisory Commission's recommendation to ban human cloning on the grounds of safety alone, arguing that issues on life and morality have been set aside. A pressing concern we face today is that the pace of our reflections in the humanities has not kept up with the fast-moving biotechnology and genetic technology. To allow human cloning too hastily when the moral, social, legal and religious issues are still unresolved is also irresponsible. His conclusion is that human cloning may well turn out to be one of the things "that we *can* do that *ought* not to be done."

Brain death is another case that highlights the conflict between modern biomedical technology and traditional culture. After a long debate, the Japanese eventually reached a compromise: the Japanese Parliament passed a law in 1998 to accept the definition of brain death in coexistence with the traditional cardiopulmonary paradigm of death, and left the room to choose to the people. But the law does not end the debate. Kazuo Takeuchi in his paper "Brain Death Criteria in Japan" describes the debate on the concept of brain death in Japan and points out that the situation there is a definite example of conflict between medical advances and the traditional culture of Japan. As for brain death, we still have two ways of thought represented by the expressions: "a dead brain in a living body" or "corpses with a good volume pulse".

However, in mainland China the concept of brain death is still not accepted in the law. Ming-Xian Shen in his paper "To Have a Good Birth as well as a Good Death: The Chinese Traditional View of Life and Its Implications" elaborates the Confucian and Taoist metaphysical and moral viewpoint of life and death and its implications for bioethics. He points out that it is relatively easy to suggest a good birth and a meaningful life, but to suggest a good, meaningful death is something quite different, because it sees the attitude toward death and the manner of death as the way of being human and the way to ideal life. He argues that there are many reasonable elements in the life-view of traditional China that are consistent with modernity and become important ideological and cultural resources of bioethics. Bioethics in Asia must have its own characteristics and should make a special contribution to global bioethics. In Ping Dong & Xiaoyan Wang's paper "Life, Death and End-of-Life Care: Taoist Perspective", it is argued that the Taoist view of life and death provides an intellectual foundation for end-of-life care, and the acceptance of a Taoist perspective on life and death will help dying people to reach the end of life's journey and satisfy an expectation of, and pursuit for, a good end with peace and dignity. A Taoist conception on life and death widens people's vision and helps them to merge their spirit with the great *tao* of a boundless universe.

Euthanasia is another major topic in Asian bioethics, and even within the same cultural context there are conflicting views. Da-Pu Shi & Lin Yu in their paper "Euthanasia Should be Legalized in China: Personal Perspective" report in detail on cases involving euthanasia and the findings of surveys conducted in mainland China. They argue for the legalization of euthanasia in mainland China and try to refute the arguments against it despite the fact that active euthanasia is illegal according to the current interpretation of Chinese Criminal Law.

4. BIOETHICS, POLICY AND LAW IN ASIA

Bioethics is the foundation of formulating, implementing and assessing policy and law involving health care. It is much less adequate or appropriate to address the policy dimension of bioethics without consideration of the socio-cultural context in which the policy is formulated and implemented. With globalization, isolationism is no longer plausible or tenable, but the critique of the policy or law in one culture from the viewpoint of another culture without consideration of the context may be inappropriate and counter-productive. The typical example concerns population policy in developing countries. Re-Feng Tang, in her paper, "Chinese Population Policy: Good Choice and Right Choice", argues that proper thinking regarding population policy should concentrate on goodness instead of rightness. She points out that Chinese thought on population policy is concentrates on goodness, while Western thought concentrates on rightness. These two ways of thinking can be ascribed to two kinds of ethics, that is, good ethics and right ethics. She argues that good ethics is more mature than right ethics, and explores the implications of good ethics on population policy with challenges to the rights talk in population policy.

China's Law on Maternal and Infant Health Care has been a controversial issue in the international genetic community. Apart from there being some flaws in this law, cultural differences between geneticists, bioethicists and lawyers from different countries cannot be ignored. Even the language of "eugenics" is so ambiguous that it bears divergent interpretations. Ren-Zong Qiu, in his paper, "Does Eugenics Exist in China? Ethical Issues in the Law on Maternal and Infant Health Care" provides some background information about the law, describes two approaches in drafting the law, and discusses ethical issues in formulating and implementing the law.

HIV/AIDS is one of the most disastrous challenges facing Asia in the twenty-first century. However, the preventive strategy in Asian countries is gravely inadequate. Yan-Guang Wang, in her paper, "AIDS, Policy and Bioethics: A New Bioethical Framework for China's HIV/AIDS Prevention" attempts to formulate a bioethical framework to deal with moral and policy issues in HIV prevention. She argues that basic principles such as nonmalfience, beneficence, respect for autonomy, and justice are not adequate for HIV/AIDS prevention, and suggests an improved bioethical framework which consists of the principles of tolerance, beneficence, autonomy and care. Within this framework, the principles of tolerance and care should play a central role. The principle of tolerance is located in the first order of this bioethical framework because the basic bioethical framework, when applied in the ordinary medical context, is based on a presumption that all patients have equal social and moral status. But when using these principles in an HIV epidemic, we find that some special social groups, such as AIDS patients, HIV+ patients, drug users, prostitutes and homosexuals are marginalized, often stigmatized and discriminated against. To view them as having equal social and moral status is very difficult for the public. Wang argues for the principle of care as improving the rational stance of the principles of nonmalficence, beneficence and respect for autonomy and as helping to solve issues within the basic bioethical framework, especially regarding how to solve special issues of the HIV/AIDS-related population.

Health care reform is another challenge facing Asia in the twenty-first century, both in developing regions and developed regions. This is true even in Hong Kong, where health care has been successfully provided, costing only 3% of GDP and providing more than 90% of the entire hospitalization services. Life expectancy is even higher than that in USA. Ho-Mun Chan, in his paper, "Justice is to be Financed Before it is to be Done: The Social Justice of the Hong Kong Public Health Care Reform", gives an overview of the health care financing system in Hong Kong, and explicates the reasons for reforming the existing financing system. He argues that the reform should be guided by the principle of social justice and that a just health care system should guarantee a decent minimal level of service for all as determined by an open and accountable mechanism of rationing. In the conclusion he evaluates the possible options for reforming the health care financing system in Hong Kong, and argues that no drastic change is needed provided that a proper rationing mechanism is in place.

In closing, I would like to express my sincere gratitude to all contributors who made such painstaking effort with this volume, and to Professor H. Tristram Engelhardt, Jr. and Dr. Ruiping Fan in particular: without the former's encouragement and the latter's assistance, this volume would never have appeared.

Institute of Philosophy, Chinese Academy of Social Sciences
Beijing, China

PART I

THE FOUNDATIONS OF BIOETHICS IN ASIA

CHAPTER 2

JULIA TAO LAI PO-WAH

CONFUCIAN AND WESTERN NOTIONS OF HUMAN NEED AND AGENCY: HEALTH CARE AND BIOMEDICAL ETHICS IN THE TWENTY-FIRST CENTURY[1]

1. HEALTH CARE AND BIOMEDICAL ETHICS IN THE TWENTIETH CENTURY: THE PRIMACY OF RIGHTS

Twentieth century Western liberal democracies have been largely dominated by two broad principles upon which the allocation and distribution of resources, particularly public resources, are based. These are the principle of right and the principle of need (see for example Wiggins, 1998; Doyal & Gough, 1991; Plant, 1990, 1986; Miller, 1976; Rawls, 1972). They provide the ethical framework for determining societal priorities, aiding professional judgment, grounding entitlements and reconciling claims. They also shape our decision-making and social legislation for public intervention, particularly in welfare choices and health care policies (Tao and Drover, 1997).

The principle of right is grounded in the notion of individual freedom. There is a strong tradition in the culture of the West which holds the view that freedom of the individual is the human essence (see for example, Hobbes, 1968; Locke, 1988; Mill, 1991; Patterson, 1991). As the central concept in Western political philosophy, the imperative of freedom is seen as natural and fundamental. It is also considered to be "the basic starting point, not only for understanding human relations and society, but for defining and justifying other concepts, such as justice, obligation, and rights" (Hirschmann, 1996, p. 55). For many of the liberal right theorists, the importance of equality is that we are equally free, and rights exist to protect spheres of action from outside interference. John Locke defends this popular assumption in Western moral and political philosophy in this way: "All men are naturally in ... a state of perfect freedom to order their actions, and dispose of their possessions and persons, as they think fit, within the bounds of the law of nature, without asking leave, or depending upon the will of any other man" (1988, p. 269). According to Locke, "The state of nature has a law of nature to govern it, which obliges everyone, and reason, which is the law, teaches all mankind, who will but consult it, that being all equal and independent, no one ought to harm another in his life, health, liberty, or possessions..." (1988, p. 271). Thus, "The natural liberty of man is to be free from any superior power on earth, and not to be under the will or legislative authority of man, but to have only the law of nature for his rule" (p. 283).

R.-Z. Qiu (Ed.), Bioethics: Asian Perspectives, 13-28.
© *2003 Kluwer Academic Publishers. Printed in the Netherlands.*

This natural liberty is not only a given of the state of nature; it is essential to humanity, a natural right. Since the time of Locke, it is widely accepted in Western liberal democracies that the state's responsibilities should be limited by individual rights based upon this naturalistic presumption of individual freedom. It is further recognized that the rights give rise to enforceable claims between citizens as well as between citizens and the state. More recently, the notion of rights has been expanded to include negative as well as positive rights (Berlin, 1971). The former defines freedom in terms of an absence of intentional interference by others. The latter defends a notion of freedom which also includes the capacity to undertake certain actions, and hence must include a number of social or welfare rights. In order to provide adequate security for individuals, many fundamental rights are embodied in laws and in international conventions. A doctrine of right has also been an important driving force behind public policy advocacy in the second half of the twentieth century. One of its major consequences is the establishment of a right to health care in modern societies, as well as the important recognition of a number of related individual rights in major health care and biomedical decisions including abortion and procreative liberty, euthanasia, and refusal of treatment (see, for example, Miller, 1981; Buchanan, 1984; Daniels, 2001; Thomson, 2001).

2. THE CRITIQUE OF RIGHTS

A principle of right is no doubt highly important to protect individuals from being swamped by the demands of the many. But there have also been serious doubts raised in recent years about the adequacy of the language of rights and about the legal system in which rights are established for reconciling claims, grounding entitlements, and resolving conflicts, particularly in bioethical and health care matters. Many of those who participate in the critique of rights, in particular some feminist philosophers, express concern about how an overemphasis on rights could obscure other important moral concerns such as power inequities, justice for particular social groups, and real divergence of cultural, religious and moral perspectives (see for example, Gilligan, 1982; Noddings, 1984; Young, 1990, Held, 1993; Tronto, 1999). In their view, "rights" are indifferent to people's actual psychological and material capacities to exercise rights, unconcerned with responsibilities except those narrowly connected with rights, and aggressively individualist, setting claimants against each other and against larger collectives.

Specifically, the concern is that the language of rights cannot provide the resources for building mutual concern and cooperative relationships between opposing parties caught in a conflict situation of competing interests. Instead of promoting discourses that focus on the negotiation of differences and the harmonization of interests, the language of rights tends to promote discourses that encourage "the contestation of rights." A classic illustration is the case of abortion described by Sarah Ruddick (1998, pp. 313-314). Her analysis shows that often in these cases, fetus' rights, first to life, and then to a minimally safe "maternal environment", are easily opposed to a pregnant woman's rights to pursue her own life plan. Moreover the father of the fetus, its grandparents and the state may also contest a pregnant woman's right to terminate or to continue her pregnancy. What becomes clear is the fact that the language of rights, and the legal system on which it is based, tend often to exaggerate rather than reduce the division between the different actors; to sabotage rather than support the creation of more inclusive and cooperative family relationships; and to deny rather than to foster the unique connection of a pregnant woman and her infant to whom she may give birth. In these and many other instances, rights could easily be turned into instruments of oppression, the voice of power and domination, instead of being instruments for the protection of the powerless.

Another major failure of the rights discourse, according to critics, is its formalistic and acontextual approach to freedom and social justice (see, for example, Hirschmann, 1995). In particular, they question the usefulness of an abstract universal principle of rights in meeting the needs of those who are confronted with concrete major life crises. Carol Gilligan (1982), for example, in her study of the right to abortion, reminds us that women referred little to their rights, but more to their relationships in their moral deliberation. It is argued that in these situations of significant life changes, the primary imperative is to identify harm, suffering and need, and to respond in effective and responsible ways to all those likely to be affected by the pregnancy. This would require a mode of understanding and empathy which is sensitive to the particularity of relationships and which a universalist notion of rights grounded in abstract freedom is unable to recognize. In similar ways, Sarah Ruddick's study of pregnant drug users (1998) shows how often the latter do not so much need rights as help. These include help to protect one's own health and the health of the fetus; help to learn the new role of a mother, and help to nurture the developing relationship.

A third criticism of the primacy of rights is raised by communitarian liberals such as Charles Taylor (1992) who question the priority of the individual and his or her rights over society. According to Taylor's arguments, asserting a right is more than issuing an injunction. There is always the presupposition that it has an essential conceptual background in some notion of the moral worth of certain properties or capacities, without which it would not make sense. Take the right to life as an example. Taylor argues that the affirmation of the right to life could never have been understood as a right just to biological non-death in a coma. To be alive now *in the meaning of the act* is to be dealing with capacities which do not simply belong to us in virtue of being alive. These are capacities which can only develop in society and in society of a certain kind. Ironically, however, these conceptual backgrounds within which one makes sense of, and gives meaning to rights, including the right to life itself is, according to Taylor, becoming increasingly lost to us because of the way the priority of individual rights and liberty are fracturing our moral world and undermining robust communities. A similar point is made by Tris Englehardt (2002, p. 32) who observes that because of the breakdown of communities, many of us will lack a narrative integration around a deep sense of self or the meaning of reality. The implication is that we will be increasingly living a life which has the appearance of "an affirmative story of self-liberation," but "which can in the end have no enduring content."

An over-emphasis on concerns with individual rights and liberty not only significantly affects what we believe about the relationship between the self and others, especially the degree to which one sees oneself as *separate* from or as *connected* to others. If taken as the primary consideration in health care policy and biomedical decisions, it can also easily lead either to an unlimited demand on government provisions to satisfy maximum individual preferences, resulting in escalating costs and questionable sustainability of the entire system, or to the gradual erosion of the collective provision of public health care because of increasing demand for freedom of choice and individual responsibility for health care through the market and private insurance. The consequence of either development will be an increasingly privatized health care system as has been happening in many quarters of the world in the last century. Either way can bring about the gradual disappearance of public health as a common good in our society, which will ironically become even more atomized with the disappearance of more and more of its common goods.

2.1. The Appeal to Basic Human Need

"Need" is the other important principle which is at the heart of welfare state provision in many Western societies in the last century. An appeal to need is common in professional practice, social policy and development planning to address issues of claim, entitlement and obligation. There have been many attempts in the literature to develop a thin notion of need which is abstract, objective and universal in order to provide a moral foundation for the non-market allocation of resources, since a thick notion is likely to be contested. One prevalent approach is to identify certain essential human capacities for grounding a general theory of human need. This has led to a widely embraced view of human agency and personhood in the West, which basically defines the essence of human agency or personhood in terms of the capacities to act rationally and to make independent and voluntary choices (see for example Buchanan, 2001, p. 137; Beauchamp, 1998). The general tenet of this view is that an entity is conceived as a person or a human if and only if it possesses certain properties similar to the following: (1) self-consciousness of oneself as existing over time; (2) ability to appreciate reasons for or against acting; and (3) ability to engage in purposive sequences of actions. Such an understanding of the essential properties of human agency or personhood is highly reminiscent of John Locke's classical analysis of a person, whom he defines as a "thinking intelligent being, that has reason and reflection, and can consider itself as itself, the same thinking thing in different times and places" (1975, II.xxvii.6). These are the more commonly accepted cognitive capacities in the philosophical literature as to what at least are some of the necessary conditions of personhood. They are conceived to be the properties which distinguish the human from the non-human animal. They are in this sense universal, cutting across cultures.

There are several objections to this view of human agency or personhood. One of the concerns is that "the degrees of personhood invited by these criteria will allow some non-human animals to be more independent and positioned at a higher level of personhood than some humans, and may even exclude some wanted humans, say some middle-stage Alzheimer's patients, from our consideration" (Beauchamp, 1998). Such a view is considered to be unacceptable not only because it may accord a higher degree of personhood to non-human animals than some demented or severely impaired patients. It raises the objection that since these are cognitive properties which have no moral significance, it is inappropriate to delineate descriptive properties of individuals in order to draw conclusions about moral standing. The fact that some being is rational and acts freely and purposively is not sufficient for establishing any form of moral standing because capacities of rationality, self-consciousness, and the like have no intrinsic connection to moral properties such as moral agency, moral judgment and moral accountability. At the same time, there are those who argue that basic needs must be understood in terms of properties that are not merely only contingently connected to being human. In order for a theory of human need to claim universality and moral force, it should be related to some notions of the pre-conditions for moral agency, moral judgment and moral accountability for which basic needs are essential for fulfilling, protecting, and advancing these moral-capacity conditions.

An alternative approach in contemporary Western theories of need is to identify survival and autonomy as the basic needs to which basic resources and services should be allocated, partly to ensure that people can function as moral agents, and partly to allow them to carry out their life plans (Plant *et al.,* 1980; Plant, 1991; Doyal & Gough, 1991; Gewirth, 1982). Autonomy is defended as a need related not to a particular morality or culture, but to the very possibility of human agency. Autonomy on this understanding implies two things: individually goal-directed behavior, and sufficient wherewithal to allow individuals to choose and deliberate. It is necessary for participation in whatever form of life is chosen by the individual person. Thus, because moral action presupposes that people are capable of deliberation and moral choice, it requires some defense of autonomy in this sense. On this basis, some scholars argue that health and autonomy constitute the most basic human needs which are the same for everyone. They further argue that all humans have a right to optimum need-satisfaction (see, for example, Doyal & Gough, 1991).

2.2. Basic Human Need and the Ideal of Autonomy

Scholars such as Richard Dworkin (1997, p. 107) have argued that autonomy is a richer notion than liberty, which is primarily conceived either as mere absence of interference or as the presence of alternatives. The notion of autonomy is tied up with the idea of being a subject, of being more than a mere passive spectator of one's desires and feelings. The central idea that underlies the concept is indicated by the etymology of the term: *autos* (self) and *nomos* (rule or law). When first applied to the Greek city state, it referred to a city which had *autonomia* when its citizens made their own laws, as opposed to being under the power of some conquering powers. The concept of autonomy therefore presupposes an ideal of the individual as an independent and reflective rational chooser who is capable of initiating action to achieve individual goals. As pointed out by Christine Di Stefano, this celebrated ideal of autonomy "captures, in an especially compelling and efficient way, the modern discovery and valuation of freedom, reason and agency housed within a conception of the self as an independent and reflexive rational chooser" (1996, p. 98).

Because such radical autonomy and freedom require a concept of "individual" as an isolated entity, free from the influence and coercion of the collective, separateness and distinctiveness are prized as constitutive attributes of an "independent" self. Individuality and human agency are created by freedom to claim the right to make self-directed rational choices. It underscores the claim expressed by Feinberg that "I am autonomous if I rule me, and no one else rules I" (1972, p. 161). It has also supported the development of a conception of individualist autonomy from which are derived the ideal of self-sufficiency and the right of self-determination (Feinberg, 1989, p. 28). On this understanding, the only legitimate connections to others are those initiated by the individual or agreed to by her, of her own free choice.

It is therefore not surprising that from this perspective of individualist autonomy, obligation can be justified only if it is taken on voluntarily, by free agreement, called "consent" or "social contract". These fundamental assumptions of "natural" freedom and equality make consent the only possible basis for obligation. The ideal relationship among human beings is the voluntary contractual relationship of consenting adults. Consent, choice and independence are the core essence of this individualist notion of autonomy. It is further used to argue for the ideal of neutrality of the liberal state in the West. The root idea is that the state must not justify its actions on the basis that some ways of life are intrinsically better than others nor should it attempt to promote some rather than others. This also explains the ascending primacy of the value of free consent in recent decades.

Such a trend is paralleled by a growing emphasis on respect for autonomy and free consent in health care and biomedical ethics in the West. For example, Bruce Miller in his much-cited article 'Autonomy and the Refusal of Life-saving Treatment' offers this core definition of autonomy: "the right to autonomy is the right to make one's own choices, and … respect for autonomy is the obligation not to interfere with the choice of another and to treat another as being capable of choosing" (1981, p. 24). The values of autonomy, individual self-determination and free consent are foundational to the development of patient-centred medicine and an autonomy-oriented biomedical ethics in many modern societies. Patient-centred medicine and autonomy-oriented biomedical ethics are committed to maximizing patient freedom of choice, patient participation in decision-making, and patient self-interest in health care. They are the important driving forces behind policy advocacy movements in support of individual right and freedom of choice in a range of contemporary biomedical issues related to euthanasia, abortion, organ transplant, and procreative liberty involving different reproductive technologies, and including commercial surrogate arrangements.

3. THE CRITIQUE OF INDIVIDUALIST AUTONOMY

Increasing respect for the autonomy of the patient has been a very salutary development in medicine to safeguard against coercive paternalism or unwarranted intervention by the state. But there are also worries that respect for autonomy has begun to eclipse all other values in health care and biomedical discussions and decisions. Increasingly there is the concern that advocates of respect for autonomy have come to promote it no longer as the prima facie primary principle of biomedical ethics, but as an absolute value that trumps all others. One problem arising from the ascending supremacy of an autonomy-based biomedical ethics is the tendency for the autonomous choice increasingly to supplant the good moral choice as the primary concern of bioethics (Engelhardt, 2002). Another major concern, as pointed out by Hardwig (1998), is that the myth of patient autonomy makes sharing dying and finding meaning in death almost unreachable to those of us living in modern society. This is because an over-emphasis on individualist autonomy has tended to marginalize family decision-making and to undermine the doctor-patient relationship, reinforcing separation and isolation rather than sharing and involvement.

In his reflections on the issue of (un)assisted suicide, Hardwig reminds us that one of the ironies of modern society is that modern medicine saves many lives and enables us to live longer. "But it thereby also enables more of us to survive longer than we are able to care for ourselves, longer than we know what to do with ourselves, longer than we even *are* ourselves" (Hardwig, 1998, p. 63). We are thus increasingly morally troubled not by deaths that come too soon, but by deaths that come too late. Patient autonomy may give us the right to make choices about our own medical treatment. Despite this autonomy, we may feel we really have no choice, largely because we are unable to find meaning in death or bring our lives to a meaningful close. We suffer from an inability to give meaning to death because death defined as an individual right and autonomous choice is isolated, disconnected, and dislodged from the web of personal relationships and community bonds that give meaning to life and death.

Most discussions of euthanasia and advance directives in the past decades have been framed almost exclusively in terms of protecting the autonomous choice of patients to decide when to die, and enshrining that freedom of choice and exercise of right in our legal system through the requirement for informed consent under the doctrine of respect for individual autonomy. Often these discussions and the law ignore the fact that people are connected and lives are intertwined. They reinforce an isolation thesis of the self which in turn promotes a kind of atomistic individualism. They are indifferent to the context of choice-making and they marginalize respect for the family (however defined) as a decision-making unit. But Hardwig argues that important decisions for those whose lives are intertwined should be made *together*, in a family discussion. Death can have meaning if it affirms our connectedness rather than our separation and isolation at a critical juncture in our life. Dying can be a growth experience, leading to enhanced individuation and a deep sense of personal meaning for the dying patient and her family, even during the last moments of life together, if it can transcend the fears of dependency, isolation and separation through sharing and caring. This is not to deny that each of us should take responsibility for herself. But it is in the sharing and caring, not in the free choosing and autonomous decision-making, that we give meaning to, and make sense of, death and dying. It is in this sense that we are truly free and autonomous in the face of death and dying.

A major difficulty of the individualist view of autonomy lies in its underlying notion that we can abstract individuals from relationships, from social context, and even from qualities of human agency that are deemed vital, namely the capacity and need for connectedness, relationships and mutual care. The exclusive conceptualization of obligation in voluntarist terms cannot explain or accommodate many aspects of social relations and bonds that our daily experience reveals. It also carries a negative evaluation of involuntary relations and unchosen obligations. Because these are viewed to have negative values, they are also regarded as weaker claims and as less deserving of public support and resources. Generally speaking, under the individualist notion of autonomy, we are obligated to do things that we would sometimes, all things considered, prefer not to do. Obligations are regarded as sources of "unfreedom", needs are viewed as deprivations and dependencies are equated with deficiencies. In short, needs are associated with the needy and with hindrance to autonomy, notwithstanding a common recognition among liberals for a right to optimum need-satisfaction.

3.1. Relational Autonomy: A Rival Model

A different approach is found in a theory of need grounded in the relational paradigm of human relations. In place of an individualist notion of autonomy predicated on a separation of human beings and a contractual paradigm of human relations which emphasizes consent and choice, some contemporary philosophers and feminist writers argue a different thesis that persons are not fundamentally separate and isolated, but rather are literally constituted by the relationships of which they are a part (see for example, Gilligan, 1982; Noddings, 1984; Held, 1993; Hirschmann, 1996). They remind us that from the moment of birth, "The autonomous self is not merely *also* connected, but is *already* connected – as an amalgam of relations with other similarly constituted selves" (Di Stefano, 1996, p. 105). Consent theory entails seeing obligation as individually negotiated contracts between otherwise unconnected individuals. This individualist emphasis is criticized as misdescribing human relations. As pointed out by Di Stefano, if relationship is the epistemological foundation, then connection is given, and obligation is a presumption of fact. That is, obligation must be considered from the standpoint of a "given," just as freedom is the given in consent theory. From this perspective, "recognition" of an obligation is not the same as "consent" to it.

Consent theory bases its claims to legitimacy on the notion that choice creates obligation; recognition, on the other hand, involves the admission of an obligation that already exists. It is primarily concerned with understanding how separate individuals can develop and sustain connections and still be separate, how they can engage in relationships and still remain free. The central approach involves asking how obligations "arise", how they come into being. But if obligation is a presumption of fact, the issue becomes not whether an obligation exists, but how it is to be fulfilled. The central question shifts to trying to determine how to strengthen connections to enhance self-fulfillment and autonomy and how to sustain relationships to support individuation and freedom.

Not surprisingly, the liberal individualist model of a separate unembedded self and the supremacy of individual consent and autonomy as a guiding moral framework for making health care decisions and for resolving bioethical issues are being increasingly called into question. Increasingly being questioned also is the conception of the human agent as an "intelligent thinking being," to whom rationality is of central importance in defining his/her personhood or human agency. Under such a conception, if rationality is lost, the value of the agent's life is lost. The logical implication is that "a brain damaged human being who lacks the relevant capacities will fail to count as a person" (Honderich, 1995, p. 656; quoted in Au, 2000, p. 213). A related implication of such an assessment is that insofar as resource prioritization is concerned, patients who are perceived as less than full persons are unlikely to get high priority in allocation. But if we were to accept that there is indeed a more important, interpersonal dimension to human agency or personhood which is defined in terms of connectedness with significant others, and is relational in nature, then evaluating human agency or personhood just by assessing the patient's functional capacity is inadequate, no matter how meticulously that assessment is conducted. This has led to an alternative claim that in assessing personhood, the more relevant consideration is perhaps whether she/he is "someone I/we can still relate to" (Au, 2000, p. 214), rather than whether "she/he is someone I/we can still regard as rational."

As Di Stefano observes (1996), we are autonomous in the sense that we have power and abilities, desires and wants and needs, but these are "relational"; they come from, exist in the context of, and having meaning only in relation to others. But to say that they are "related" should not be taken to mean that they are "one". Such a perspective can prompt us to search for answers to health care and biomedical questions which are more oriented toward creating mutual trust among actors, building identity, achieving security and sustaining interdependency. From this perspective, the independent and autonomous agent of liberal theory is an illusion. Autonomy can only develop and thrive on the basis of responsive relational ties and community support.

3.2. Confucian Relational Self

Increasingly there is also growing recognition that the liberal individualist model of autonomy and human agency is not universally applicable to all cultures. In many Eastern societies, Confucianism is an important moral and intellectual resource which offers a different self-understanding to the liberal model of the human agent. Confucian moral philosophy also presupposes human relatedness, rather than separation of persons, as the essence of human existence. The Chinese Confucian conception of human agency is essentially non-individualistic, non-contractual and relational in nature (Tao, 1998, pp. 578-579). Care, compassion and responsibility to others are the moral capacities that define us as human. This has led some scholars to observe that in the Chinese Confucian tradition, the human subject is never seen as an isolated individual but is always conceived of as part of a network of relations (King, 1985; Tu, 1985). A person is always a "person-in-relations" – a point well captured by Wu, who writes: "Traditionally, a Chinese seldom thought of himself as an isolated entity. He was a concrete individual person who moved, lived and *had his being* in the natural milieu of the family..." (1967, p. 340).

In emphasizing the importance of human relationships, Mencius (?372–?289 B.C.), one of the most influential followers of Confucius (551-478 B.C.), also delineates this distinction between humans and brutes: "Slight is the difference between man and the brutes. The common person loses this distinguishing feature, while the gentleman (benevolent or moral person) retains it. Shun (sage king) understood the way of things and had a keen insight into human relationships. He followed the path of morality" (*Mencius* 4A:19).

To the Confucian Chinese, humans are distinguished from animals by possessing the capacity to follow morality in forming distinctive kinds of relationships with other human beings to realize their human nature. What is therefore crucial for humans is to be self-consciously aware of "a keen insight" in human relationships and to follow "the path of morality," the final destination of which is to exhibit perfect human relations as did the ancient sages. This is further explained by Mencius when he says: "The [carpenter's] compass and square produce perfect circles and squares. By the sages, the human relations are perfectly exhibited" (*Mencius* 4A:2).

Exhibiting perfect human relations is the goal of human development which is just as "natural" as it is the goal of compass and square to produce perfect circles and squares. Thus, to be in association with others is fundamental to the Confucian conception of the human subject. It is only in the midst of one's social relations that one learns to be human and realizes one's moral nature. The moral starting point of the Confucian self is relationship with others and not individual freedom and rights. It is in one's role relationships and role performance that the self finds the source of one's sanctity as a human being. In this way, Confucian self-understanding emphasizes a constant process of transformation through role relationships and obligations (Tao, 1995, p. 16). Through role relationships, one experiences one's bonding with others; through role obligations, one realizes one's individuality. On the Confucian conception, neither "self" nor "other" is a means to the other's end, yet at the same time, each is absolutely necessary to the other for its realization and fulfillment as truly human.

In mainstream Western liberalism, the notion of human subjects as equal distinct entities being prior to, and independent of, experience is a necessary requirement of moral agency. Within the liberal self-conception, individual persons are free and equal agents. Society arises out of the voluntary agreement among independent rational contractors for mutual gains and peaceful co-existence. In contrast, there is no notion of human subjects as equal distinct entities being prior to, and independent of, experience in Chinese Confucian moral and political philosophy. As pointed out earlier, the Confucian conception of society is non-contractual. In a contractual society, human relations are regulated by the concept of individual rights. In the Confucian non-contractual society, human relations are guided by the concept of *jen* or benevolence (Tao, 1998, pp. 606-607). Individuals are bound together by relations of concern and caring, empathy and reciprocity. They are not only deeply affected by their obligations and relations with others, they are partly constituted by these relations and obligations.

3.2.1. Confucian Relational Autonomy

It is important to note that Confucian teachings also attach great autonomy to the individual self. True, the Confucian self is not an atomic isolated entity, but stands in interdependence with its social environment. But the human subject in the Confucian tradition is not essentially or even exclusively defined by the social role. In many of the sayings of Confucius and Mencius, we learn that the Confucian protagonist is positively a "moral subject" in the sense of a self-responsible autonomous being with his or her own independent will and dignity. The *Analects* (the collected sayings of Confucius) describes the concept of a "self" that stands its ground against the misunderstandings and the hostility of others. It can be self-critical and sits in judgment on itself. This faith in the autonomy of the human will is borne out by one of the most quoted sayings by Confucius in the *Analects* (9:26), where he reminds us that "It is possible to deprive a whole army of its generals. But it is not possible to deprive even a commoner of his/her will." From Mencius, we also learn that the human subject is not only capable of critical self-reflection, but is also someone who is able to "withstand thousands or tens of thousands" if convinced that he has been upright in his behavior, and who "strides his way alone," if necessary, and cannot be "led astray by poverty and mean conditions," or "bent by authority and power" (*Mencius* 3B:6). The Confucian belief in the human mind's autonomy to choose and reject according to its own judgement can hardly be expressed more clearly. But Confucianism poses a distinction between individualism and autonomy. It emphasizes human connectedness rather than separateness of persons. Unlike the liberal individualist tradition, the moral person in the Confucian project is not only self-directed, but is also other-directed.

Reciprocity, not equality, is the central principle which structures interactions and interpersonal relationships in the Confucian moral tradition. The notion of reciprocity has many interpretations. One approach is to conceive of it as a give-and-take relationship emphasizing a simple tit-for-tat notion, referring to direct and exact return in kind. In the Confucian account, reciprocity is not conceived in terms of a kind of transaction like an exchange of gifts or goods. The moral basis of reciprocity is not a social contract, but rather our interconnectedness and interdependence (Tao, 1999, p. 579). It involves, on the one hand, giving without consideration of return or obligation, and on the other hand, in the words of Lawrence Becker, a straight moral obligation on the part of the recipient "to return good for good" (1986, p. 3). Understood in this way, reciprocity is a recipient's virtue.

As a procedural rule in human relationships, reciprocity is expressed, in its negative form, through the Confucian doctrine of "*shu*" which is explained as "Do not do to others what you do not want them to do to you" (*Analects*, 15:28). The positive form of reciprocity is expressed through the doctrine of "*Zhong*" according to which an agent, "in wishing to establish himself, will also establish others. And wishing to achieve perfection, he will also perfect others" (6:30).[2] The requirement to follow the rule of reciprocity in thinking about moral responsibility enables the agent to see herself as part of a larger overlapping network of reciprocal relationships, enabling her also to recognize even non-voluntary obligations – obligations incurred in the course of social life without regard to consent, invitation, or acceptance. For the Confucians, standing on one's own feet does not mean standing apart from other people. The web of reciprocal obligations or moral relations in which one finds oneself, defines oneself. In a truly reciprocal role relationship where there is mutuality in the interaction, "self" and "other" are both constituted as well as constitutive of each other in the bonding and individuation which take place within the relationship. The reciprocity of benevolence in relationship is the guiding principle which has structured society and human interaction in China for nearly two thousand years.

3.3 Connected Selves and Relational Needs: Paradigm for a New Biomedical Ethics

The Confucian relational view that individuals are mutually self-creating within relationships lends support to the feminist critique of the liberal model of the separate self by emphasizing the importance of the claims of human connectedness and particularistic relationships in moral development and ethical decision making. It can provide a corrective perspective to excessive individualism, without losing sight of the importance of the autonomy of the individual. Specifically, it supports a relational view of human needs which understands needs as arising from a connection of relationship between two or more participants moving towards differentiation and individuation. On this understanding, basic human needs are conceived not merely as needs for individual autonomy, choice and independence.

Basic human needs are also needs for care, bonding and interdependency. This view differs from the individualist view which tends to conceptualize needs in terms of some properties of the human being as individual, which may be reason, desires, rights, etc. The Confucian tradition stresses a different moral starting point which is relational in nature. It provides insights for the integration of the basic tenets of the individualist and the relational paradigms of human needs in the West, summarized in the following way (see Manning, 1998; West, 1992; Chan, forthcoming, 2004):

<u>Individualist and Relational Conceptions of Human Need and Agency</u>

	Individualist Paradigm	Relational Paradigm
Self Understanding	Separate, Unembedded Self	Connected, Embedded Self
Values	Freedom Self-determination Self-sufficiency	Connection Reciprocity Responsiveness
Fears	Attachment Paternalism Dependency	Separation Isolation Abandonment
Human Need and Agency	Autonomy Choice Independence	Care Bonding Interdependence

Chinese Confucianism shares a similar emphasis with Western feminist care ethics on the centrality of human relationships to our happiness and well-being. It allows us to recognize dependency relations as foundational human relations and to acknowledge that dependents whose intelligence or independence might have become severely compromised, as in the case of Alzheimer's patients or demented elderly relatives, are still importantly contributing to the ongoing nature of human relationships even in their utmost neediness. The relevant criterion for making health care and biomedical decisions is not so much whether the patient or person has lost his/her rational capacities, but rather whether "she or he is someone who I/we can still relate to." Such a conception enables us to focus our social intervention on the importance of the central human bonds that form around dependency needs and to protect and enhance these bonds and ties which are central to our very identities. It also makes possible a different conception of autonomy. Whereas the liberal individualist notion of autonomy takes it as a given, basic human need which is a precondition of human agency, the relational paradigm conceives of autonomy as a life-long development task in a mutually self-creating, interdependent relationship among participants (Doyle and Gough, 1991).

4. CONCLUSION

The excessive emphasis on a patient-centered, autonomy-oriented approach to bio-medical ethics in the last century has indeed failed to take into sufficient account the context of choice-making and the fact that the interests of family members are closely connected (Chan, forthcoming, 2004). Bio-medical ethics in the twenty-first century should try to move beyond the myth and the supremacy of liberal autonomy to give full recognition and respect to the relational nature of human need and agency. It should be guided by a moral framework which is not promoting isolation and reinforcing a kind of atomistic individualism. Health care and biomedical discussions and decisions should be conducted within a moral framework which aims to create and sustain relationships of mutual care, interdependency and responsibility to foster growth, development and individuation.

To some extent, one can say that under the liberal individualist paradigm, the primacy of rights to the patient is well-assumed. Under the relational paradigm, the primacy of the family to the patient is well-presupposed. A relational paradigm of human need and agency, which constitutes the basic tenet of a feminist care ethic and Confucian moral philosophy, better allows us to talk about the relationship between doctors and the families of patients, and to bring patients' family relations into the orbit of care and concern. The emphasis on respect for family as a decision-making unit disallows us from treating the patient as an isolated independent being, constituted by his rationality and autonomy rather than by his ties and attachments. This is of course not to suggest that either one of them can be reduced to the other. A Confucian or feminist relational approach to bioethics reminds us that the value of the human agent is not grounded in rationality or autonomous choice-making alone, but in meaningful relationships of interdependency and human interactions.

The Confucian or feminist recognition of the relational source of the self enables us to hold the view that society consists not of isolated individuals but of relationships, and that it is essentially held together by human connections and not through a system of abstract rules or contracts. An awareness of these bonds between people leads to the further recognition of our mutual responsibility for one another and to insights into the necessity of sympathy and care to provide a more adequate response to human needs in biomedical decisions and choice-making in the twenty-first century. To see rights and autonomy from an individualist, unattached point of view is to see them as creating a private sphere of interests and a personal bundle of entitlements for private consumption. Seeing rights and autonomy from a relational, engaged point of view is to see them as creating a public sphere of free discourse and sustaining a bundle of public institutions for the public good, including the good of medicine.

Starting from an assumption of connection and obligation allows for many political debates on public policy issues regarding abortion, birth control, reproductive technology, financial assistance to care givers or treatment of brain-damaged patients. It also opens up our inquiry to see that the notions of individuality and freedom that underlie autonomy are rooted in contexts of relationships. Importantly, it leads us to develop a relational view of autonomy which can be used as an important conceptual resource to enable us to see "interdependency," and to acquire a notion of self as deeply situated in relationships.

As a matter of fact, in many societies in the East, health care choices and biomedical decisions are much guided by a self-understanding which emphasizes that self and others are interdependent, and that the self is by definition connected to others. In Japan, for example, the law requires that organ transplant should only be performed where there is the donor's living will permitting donation, with the donor card specifying the organ and with the family's consent (Yamada, 1999). In a study of the Hong Kong Chinese perception of human need conducted by the author[3] (Tao et al., 1996), the findings also show that there is strong endorsement of a relational understanding of the self as opposed to an individualist view among the Hong Kong Chinese. There is much greater emphasis on connectedness, social context and relationships in their view of the self of the human agent. They also tend to give higher priority to care, bonding and interdependency as important basic human needs, as compared with the more individualist needs of autonomy, choice and independence. The study shows that a relational paradigm of human agency and autonomy is not specific to contemporary feminist thinking or critics of Western liberal individualism.

The philosophical insight of a relational notion of autonomy analyzed in this paper enables us to see the centrality of human relationships to our human agency, and to recognize dependency relations, even those unchosen and involuntary, as foundational human relations. It can help us focus on the importance of the central human bonds that form around dependency needs, and to appreciate dependents, who, even in their utmost neediness, or severe loss of rational capacity, contribute to the ongoing nature of human relationships which are a value in themselves.

Asia Research Centre
and
Department of Public and Social Administration
City University of Hong Kong, Hong Kong, PRC

NOTES

1. The author wishes to register a note of deep appreciation to the Centre for Modern Chinese Studies (philosophy project) and to St. Hugh's College at University of Oxford, UK, where I was offered visiting fellowships during the period of my sabbatical leave from the City University of Hong Kong from April 1999 to December 1999 where I developed and completed this paper under their invaluable support.
2. Chan See Yee offers a different interpretation in "Disputes on the *One* Thread of *Chung-Shu*", in *Journal of Chinese Philosophy* 26:2 (June) 165-186. She is of the view that *chung* is an attitude of exerting one's best effort to serve others, while *shu* is a directive for moral actions. In other words, *chung* is the spirit which underlies the practice of *shu*.
3. The author also wishes to express gratitude to the Research Grants Council of the University Grants Council in Hong Kong for a Competitive Earmarked Research Grant (CERG) to conduct the study on "Needs and Welfare Choices in Hong Kong" in 1996 in association with Glenn Drover, Wong Chack-kie and Kenneth Chau (RSC/RGC Cityu 905/96H). The study provided invaluable theoretical insights for grounding the conceptual analysis of different notions of need argued in this paper. A much shorter version of this paper was presented to the Australasian Society for Asian and Comparative Philosophy Conference *Asian and Comparative Philosophy: Philosophy and Culture*, Sept. 30-Oct. 3, 1998, at the University of New South Wales, Sydney, Australia.

REFERENCES

Au, Derrick K.S. (2000). 'Brain injury, brain degeneration, and loss of personhood.' In: Gerhold K. Becker (Ed.), *The Moral Status of Persons: Perspectives on Bioethics* (pp. 209-218). Amsterdam: Athanta.

Beauchamp, Tom (1998). 'Personhood, autonomy, and rights.' Paper presented at the *International Symposium on Bioethics and the Concept of Person.* Centre for Applied Ethics, Hong Kong Baptist University.

Becker, Lawrence (1986). *Reciprocity.* London: Routlege & Kegan Paul.

Berlin, Isaiah (1971). 'Two concepts of liberty.' In I. Berlin (Ed.), *Four Essays on Liberty.* New York: Oxford University Press.

Buchanan, Allen (1984). 'The right to a decent minimum of health care,' *Philosophy and Public Affairs.* 13, 55-78.

Buchanan, Allen (2001). 'Advance directives and the personal identity problem.' In: John Harris (Ed.)., *Bioethics* (pp. 131-156). Oxford: Oxford University Press.

Chan, Ho-Mun (forthcoming, 2004). 'Sharing death and dying, advance directives, autonomy and the family.'*Bioethics.*

Daniels, Norman (2001). 'Health-care needs and distributive justice' In: John Harris (Ed.), *Bioethics* (pp. 319-346). Cambridge: Oxford University Press.

di Stefano, Christine (1996). 'Autonomy is the light of difference.' In: Nancy J. Hirschmann & Christine di Stefano (Eds.), *Revisioning the Political* (95-116). Boulder: Westview Press.

Doyal, Len & Gough, Ian (1991). *A Theory of Needs.* London: Macmillan.

Dworkin, Gerald (1997). *The Theory and Practice of Autonomy.* Cambridge: Cambridge University Press.

Englehardt, H. Tristram Jr. (2002). 'Morality, universality, and particularity: Rethinking the role of community in the foundations of bioethics.' In: Julia Tao Lai Po-wah (Ed.), *Cross-Cultural Perspectives on the (Im)Possibility of Global Bioethics* (pp. 19-40). Dordrecht: Kluwer Academic Publishers.

Feinberg, Joel (1972). 'The idea of a free man.' In: R.F. Dearden (Ed.), *Education and the Development of Freedom* (pp. 3-29). London: Routledge & Kegan Paul.

Feinberg, Joel (1989). 'Autonomy.' In: John Christman (Ed.), *The Inner Citadel: Essays on Individual Autonomy.* New York: Oxford University Press.

Gewirth, Alan (1982). *Human Rights: Essays on Justification and Application.* Chicago: Chicago University Press

Gilligan, Carol (1982). *In a Different Voice.* Cambridge, Mass.: Harvard University Press.

Grimshaw, Jean (1988). 'Autonomy and identity in feminist thinking.' In: Morwenna Griffiths & Margaret Whitford (Eds.), *Feminist Perspectives in Philosophy* (pp. 90-108). Bloomington: Indiana University Press.

Hardwig, John (1998). 'Dying at the right time: Reflections on (un)assisted suicide.' In: Hugh LaFollette (Ed.), *Ethics in Practice: An Anthology* (pp. 53-65). Massachusetts: Blackwell.

Held, Virginia (1993). *Feminist Morality: Transforming Culture, Society and Politics.* Chicago: Chicago University Press

Honderich, Ted (Ed.) (1995). *The Oxford Companion to Philosophy.* Oxford: Oxford University Press.

Hirschmann, Nancy J. (1996). 'Revisioning freedom: Relationship, context, and the politics of empowerment.' In: Nancy J. Hirschmann & Christine Di Stefano (Eds.), *Revisioning the Political* (pp. 51-74). Boulder: Westview Press.

Hobbes, Thomas (1968). *Leviathan.* C.B. Macpherson (Ed.). New York: Penguin.

King, Ambrose (1985). 'The individual and group in Confucianism: A relational perspective. In: Donald Munro (Ed.), *Individualism and Holism: Studies in Confucian and Taoist Values* (pp. 57-70). University of Michigan: Ann Arbor Centre for Chinese Studies.

Lau, D.C. (Trans.) (1979). *Confucius: The Analects.* London: Penguin.

Legge, James (Trans.) (1960). *The Chinese Classics. Volume II: The Works of Mencius.* Hong Kong: Hong Kong University Press

Locke, John (1988). *Two Treatises of Government.* Peter Laslett (Ed.). Cambridge:Cambridge University Press.

Lock, John (1975). *An Essay Concerning Human Understanding.* Peter Nidditch (Ed., with an Introduction, Critical Apparatus and Glossary). Oxford: Clarendon Press.

Manning, Rita C. (1998). 'A care approach.' In: Helga Kuhse & Peter Singer (Eds.), *A Companion to Bioethics* (pp. 98-105). Oxford: Blackwell.

Mill, John Stuart (1991). *On Liberty.* New York: Oxford University Press.

Miller, Bruce (1981). 'Autonomy and the refusal of lifesaving treatment,' *The Hastings Center Report* 22(9), 24
Miller, David (1976). *Social Justice.* Oxford: Clarendon Press.
Noddings, Nel (1984). *Caring: A Feminist Approach to Ethics and Moral Education.* Berkeley: University of California Press
Patterson, Orlando (1991). *Freedom: Freedom in the Making of Western Culture.* New York: Basic Books.
Rawls, John (1972). *A Theory of Justice.* Oxford: Oxford University Press.
Plant, Raymond, Lesser, H. & Taylor-Gooby, Peter (1980). *Political Philosophy and Social Welfare.* London: Routledge
Plant, Raymond (1990). 'Citizenship and rights.' In: IEA (ed.), *Citizenship and Rights in Thatcher's Britain: Two Views.* London: Institute of Economic Affairs.
Plant, Raymond (1986). 'Needs, agency and rights.' In: C. Sampford & J. Law (Eds.), *Law, Rights and the Welfare State.* London: Croom Helm.
Plant, Raymond (1991). *Modern Political Thought.* Oxford: Blackwell.
Ruddick, Sara (1998). 'Reproductive ethics.' In Qiu Renzong, Jin Yihong & Wan Yanguang (Eds.), *Chinese Women and Feminist Thought.* Beijing: China Social Science Publishing House.
Tao, Julia (1999). 'Does it really care? A critique of the Harvard Report for Health Care Reform in Hong Kong,' *Journal of Medicine and Philosophy*, 24(6), 571-590.
Tao, Julia (1998). 'Confucianism.' In: Ruth Chadwick (Ed.), *Encyclopedia of Applied Ethics* (pp. 597-608). San Diego: Academic Press.
Tao, Julia (1995). 'The moral foundation of welfare in Chinese society: Between virtues and rights.' In: Gerhold Becker (Ed.), *Ethics in Business and Society: Chinese and Western Perspectives* (pp. 9-24). Heidelberg: Springer-Verlag.
Tao, Julia & Drover, Glenn (1997). 'Chinese and Western notions of need,' *Critical Social Policy* 17(1), 5-25.
Taylor, Charles (1989). 'Cross purposes: The liberal-communitarian debate.' In: Nancy Rosenblum (Ed.), *Liberalism and the Moral Life.* Cambridge: Harvard University Press.
Thomson, Judith Jarvis (2001). 'A defence of abortion.' In: John Harris (Ed.), *Bioethics* (pp. 25-41). Cambridge: Oxford University Press.
Tronto, Joan (1999). *Moral Boundaries: A Political Argument for an Ethic of Care.* New York: Routledge.
Weiming, Tu (1985). 'Selfhood and otherness: The father-son relationship in Confucian thought.' In: *Confucian Thought: Selfhood as a Creative Transformation* (pp. 113-130). New York: New York University Press.
West, Robin (1992). 'Political theory and gender.' Reprinted in: John Arthur & William H. Shaw (Eds.), *Social and Political Philosophy* (pp. 569-593). Engelwood: NJ. Prentice Hall.
Wiggins, David (1998). *Needs, Values, Truth.* Oxford: Clarendon Press.
Wu, John C.H. (1967). 'The status of the individual in the political and legal tradition of old and new China.' In: Charles Moore (Ed.), *The Chinese Mind* (pp. 340-364). Honolulu: University of Hawaii Press.
Yamada, Takao (1999). 'Recent development of bioethics in Japanese style.' Paper presented at the International Conference on Bioethics *Individual, Community & Society: Bioethics in the Third Millennium*, Centre for Comparative Public Management and Social Policy, City University of Hong Kong.
Yee, Chan See (1999). 'Disputes on the *One thread* of *Chung-Shu*," *Journal of Chinese Philosophy* 26(2), 165-186.
Young, Iris (1990). *Justice and the Politics of Difference.* Princeton, N.J.: Princeton University Press.

CHAPTER 3

EDWIN C. HUI

PERSONHOOD AND BIOETHICS: A CHINESE PERSPECTIVE

1. SUBSTANCE, BEING, AND PERSON

In the West, there is a long tradition of thinking of "being", and hence of thinking of the "person" in terms of "substance". The conception of substance as the primary instance of real being, and hence of the person, can be traced back to the thought of Aristotle, if not Plato or even further back to the ancient pre-Socratic Greeks. One of the most crucial characteristics of the concept of substance is its aptitude of existing in itself as a concrete individual thing and not as a part of any other being. This criterion of substance Aristotle has called "the most distinctive mark of substance". As a primary instance of real being, this power to exist *in itself* as an ultimate subject of action and attribution, and not as a part of any other being, is perhaps the most important thing one can say about substance. This emphasis of the "in itselfness" of the substance is necessary, because wherever there is real being, there must be substance to ground whatever else is there as the being's ultimate subject of inherence. Without the "in itselfness" of the substance, every instance of real being could be a part of, or inhere in, some other being, and then the necessary fulfilling conditions for any being to exist would be infinitely deferred and theoretically never fulfilled. Substance, with its "in itself" characteristic, is necessary in making a being to be what Bernard Lonergan has called a "unity, identity, whole".

2. RATIONAL SUBSTANTIAL HUMANITY OF THE MODERN WEST

Philosophical circles of the modern West have largely retained this "substantialist" tradition, although there are various and divergent expressions of the "being" and "person" of humanity. Some favor a material view and consequently define human personhood in terms of the body, genome or the development of the human brain. Others hold a moral view of personhood and define personhood as an entity who has rights. And those with a religious view regard the human person as one who has been given a soul by God. But since the Enlightenment, the most prevalent view has been to define a person as a rational being with consciousness and self-consciousness. This "modern rational and substantial humanity" is best encapsulated in the thought of René Descartes (1596-1650), the "father of modern thinking," who formulated the "self" as *"cogito ergo sum"* - "I think therefore I am." As Descartes remarks, "I noticed that, whilst I thus wished to think all things false, it was absolutely essential that the I who thought this should be something" (1995, p. 156).

R.-Z. Qiu (Ed.), Bioethics: Asian Perspectives, 29-43.
© *2003 Kluwer Academic Publishers. Printed in the Netherlands.*

Descartes considers this truth the first principle of philosophy which is "so certain and so assured that all the most extravagant suppositions brought forward by the skeptics were incapable of shaking it..." (p. 156). On this basis, he concludes what he judges to be a human being: a thing that thinks, i.e., a mind, a mental substance (which he calls soul in order to make it intelligible to the audience of his time).

Furthermore, in Descartes' view, this mental substance, the mind, does not depend on the body to exist, and hence the former is not only different from but also independent of the latter. It continues to exist even when the body no longer exists, and hence its immateriality provides the basis for personal immortality. In Descartes' own words:

> I knew that I was a substance the whole essence or nature of which is to think, and that for its existence there is no need of any place, nor does it depend on any material thing; so that this "me", that is to say, the soul by which I am what I am, is entirely distinct from its body, and is even more easy to know than is the latter; and even if the body were not, the soul would not cease to be what it is (p. 156).

By presupposing the "I" as a substance in which thinking is realized (i.e., thinking itself presupposes the existence of the substance "self" or the self-conscious "I"), Descartes' formulation "I think therefore I am" affirms to him his own existence as a thinking self, but it proves little beyond that. If I am a thing which thinks, what is a thing which thinks? While he is right in retaining the concept of the "self" as substance, his new-found formula "cogito ergo sum" conceives of the "self" as a thinking substance which has no need of the world for its existence, subsisting in its own consciousness, with all the eternal truths innate in it. He declares he can have no external knowledge except by means of the ideas he has within him. In this tradition, person is defined as an ego subject subsisting in itself, endowed with reason in his inner being. In short, a person is a rational individual substance.

3. INADEQUACIES OF THE HIGHER-BRAIN DEFINITION OF PERSONHOOD

Modern bioethical discourse in the West has shown a clear proclivity to follow the Cartesian framework and to define human personhood in terms of higher-brain functions or some similar psychological criteria. For example, Joseph Fletcher identifies the four essential traits of personhood as neocortical function, self-awareness, relational ability and happiness (1975, pp. 4-7). Michael Tooley argues that a being must possess self-consciousness, i.e., the awareness of the self as a continuing subject of experiences and other mental states, before personhood can be said to be present. Without the capacity of self-awareness, Tooley reasons, there can be no desire to continue to live; such an entity does not possess the right to life and hence is not a person (1973, pp. 51-91). Building upon a similar psychological foundation of personhood, Daniel C. Dennett proposes six conditions for personhood: rationality, consciousness, self-consciousness, ability to treat and to be treated with an attitude befitting a person and the ability to engage in verbal communication (1976). H. Tristram Engelhardt Jr. also advocates rationality, self-consciousness and a minimal moral sense as the most important conditions for what he calls "strict personhood" which endows individuals with rights and duties (1996, pp. 135-140). Perhaps the most remarkable criterion for personhood which is based on one's psychological faculties is that proposed by Harry G. Frankfurt, which he calls "second-order volitions." To qualify as a human person, Frankfurt contends that not only one must be capable of having desires (itself an ability which presupposes some forms of rationality, consciousness and perhaps some degrees of self-consciousness), one also must possess the ability to counter such a primary or "first order" desire. In other words, one must be able to desire not to desire—a "second order" desire. Finally, one must possess the freedom of the will to act on the "second order" desire. When a human individual demonstrates this type of "second-order volitions," then one is truly a person (Franfkurt, 1971, pp. 15-20).

The inadequacy of this "rationalist' approach is often acknowledged by its own proponents, and many feel the need to create separate provisions to include human beings who exist in the margins of life. Charles Hartshorne, for example, is adamant that only human beings with the capacity to speak, reason and make decisions are qualified to be called persons. These are "actual persons" while fetuses and the hopelessly senile are only "possible persons." Yet by his own admission, some of these "possible persons" may have to be treated as persons for "symbolic reasons" (1981, pp. 42-45). Another writer, James W. Walters, also believes that only "individuals who possess certain capacities of the higher brain inherently lay down to the designation of person" (1997, p. 26), yet he goes on to criticize those who embrace this position (presumably including himself) as extreme, and argues that this position "fail[s] to take their positions to strict logical conclusions" (p. 62) and therefore is in the end not adequate. Walters admits that while the personhood (i.e., higher-brain) perspective is appealing, "its inadequacy is particularly apparent in dealing with questions of marginal life..." (p. 63). He goes on to provide a much more nuanced account of what he terms "proximate personhood" which takes into account other factors including human potentiality, physical development and social bonding (pp. 62-77). Engelhardt basically also embraces the higher-brain capacity as his criterion for personhood, but he also recognizes that people treat and respect human beings who do not yet possess or have lost their rational capacities as persons. So while he posits "strict persons" as those who possess rational capacities, he allows neonates, infants, the senile and profoundly retarded individuals as persons in a social sense. In other words, their status of personhood is imputed on the basis of social considerations (1996, pp. 146-153). What these writers are saying seems to be that while the psychological definition must be maintained as the reigning definition of personhood, many human beings have been left out. And because the dominant social ethos does not permit them to be "defined out of existence" – infants cannot be killed and those in the permanent vegetative state cannot be buried or have their organs removed for transplant – somehow other provisos have to be made available to include them as persons.

The ambivalence exhibited by these advocates of the higher-brain definition of personhood suggests that such a definition is fundamentally deficient in its description of human nature. As we shall see in what follows, a Confucian account of personhood emphasizes both the "substance" dimension and the importance of "relationality" in the constitution of personhood, and serves to supplement the "rationalist" account of personhood in the West. For one thing, such a narrowly "rationally" construed definition has overlooked other dimensions of the human psyche such as one's memory, imagination, emotion and subconsciousness which not only are integral aspects of a person, but also play a crucial role in contributing towards one of the most important dimensions of human nature, namely its self-transcending capacity. Here, self-transcendence is taken to be the sense that the self is not only capable of standing apart from itself to reflect on itself but also reaching out to others toward mutuality and community. Without taking into consideration one's emotional life and subconscious mental world, an individual's inner unity cannot be maintained. Inner conflicts or imbalances often result in such an individual and manifest themselves as an inability or reluctance to open in response to others. It may also incline an individual to be rigidly enclosed in self-absorption and self-encapsulation, resulting in a narcissistic life totally alienated from others. In such a state, an individual's freedom is curtailed, diminished or lost, since freedom includes one's possibility not only to exercise self-determination in choice, but also in response to self-transcendence toward the world beyond the self. The truly authentic free self is one which is neither imprisoned by nor burdened with itself, but freely, creatively and responsively makes itself available to others. This mandates that we take seriously the social and relational basis of personhood.

4. THE PERSON AS A PSYCHOSOMATIC UNITY

Traditional Chinese philosophy recognizes that human life is made up of two essential components: "*xing*" or "*ti*" (form, body) and "*shen*" (psyche, spirit) and is concerned with the way the two are related to each other. Those who are more inclined toward naturalism, e.g., philosophic Taoists, prefer to see the spirit as dependent on the body; and those with theistic beliefs, e.g., religious Taoists, hold the view that the body is dependent on and given life by the spirit. A third option suggests that the spirit and the body co-exist in the human person but are independent of each other. Yet all three approaches agree that both the body and the spirit are constitutive of human life and derived from "*qi*" (vital force). In a passage from the book *Kuan Tze* attributed to the first Legalist Kuan Chung (d.645 BC) it is stated that "in the birth of a human being, the heaven provides the psyche, and the earth supplies the body. When the two are united, a human being is formed. When the two elements (psyche and body) are in harmony, the person lives; but if they are in disharmony, the person will not live" (*Kuan Tze*, Bk XVI, Sec. 49). Even though this teaching introduces a strong element of dualism in the understanding of human nature, it nevertheless seeks to underscore both the physical and psychological dimensions of human personhood. Apparently, it does not matter much which one of the two entities has priority; what matters is the fact that they exist in a harmonious psychosomatic unity.

Another important reason why the body is accorded with a high place in Confucian thought is due to a second level of understanding of holistic personhood which relates Confucian anthropology to the Confucian conception of an ideal person as a moral sage. For example, Mencius taught that a sage is a person whose moral character has been perfected. Since no one is born a sage, to become a sage means that the person is a "being" always in the process of "becoming." This implies that all the three aspects of the person - the psyche (*shen*), the vital force (*qi*), and the physical body (*ti*) - are imperfect, waiting to be perfected. This can be seen in three passages from Mencius. For the heart, Mencius said, "[f]or a man to give full realization to his heart is for him to understand his own nature, and a man who knows his own nature will know heaven" (*Mencius* 7A:1; Lau, p. 265). In other words, self-knowledge is the means to progress toward perfection of one's mind, and only a few are able to attain perfect self-knowledge in their life-time. As for *qi*, the vital force, Mencius said, "if the will is concentrated, the vital force [will follow] and become active. If the vital force is concentrated, the will [will follow it] and become active... And I am skillful in nourishing my strong, moving power [force, *qi*]" (*Mencius* 2A:2; Lau, p. 55). To nourish one's *qi* is one of the most important lessons of self-cultivation in Confucian thinking. And finally for the body, Mencius said, "[o]ur body and complexion are given to us by Heaven. Only a sage can give his body complete fulfilment" (*Mencius* 7A: 38). This passage is interpreted to mean that it takes the moral character of a sage to actualize and fulfill the body; in turn, Mencius taught that the qualities of sagehood are exhibited through the body. "That which a gentleman follows as his nature, that is to say, benevolence, rightness, the rites (propriety) and wisdom, is rooted in his heart, and manifests itself in his face, giving it a sleek appearance. It also shows in his back and extends to his limbs, rendering their message intelligible words" (*Mencius* 7A: 21; Lau, p. 271). So, the whole body manifests the glory of the moral person.

This high regard of the human person as a psychosomatic unity has led traditional Chinese medicine to put a heavy emphasis on the interactions and functions of the physical and psychical elements in the human person. For example, it is believed that the secret to avoid disease and illness is to ensure the unimpeded circulation of the "psychic energy" in the body. Also it is believed that different "psychic energies" in the various organs of the body must be preserved if health is to be maintained. From the second century BC on up to about the seventeenth century AD, Chinese physicians continued to believe that the energies associated with the five psychological faculties, spirit, will, soul, animal spirit and ideas, take abode in and are controlled by the five organs, heart, kidney, liver, lung, spleen respectively (Huang Ti, p. 23). When the five organs are protected, the psychic energies are safeguarded, and the mental faculties will be preserved. The notion that one's essential psychical forces are found inside bodily organs has a major impact on medical ethics, especially in the area of surgery and organ transplant. It explains partly why Chinese people are generally resistant to surgical intervention, for they fear that once the bodily cavities such as the abdomen or the chest are surgically opened, the various forces or energies will escape out of the body. It also explains why the Chinese are generally reluctant to donate organs for transplant, for they fear that the psychological or spiritual elements which reside in the organs may be given away when they may have post-mortem significance. A psychosomatic view of the human person also means that when a Chinese person has lost his mental faculty he may consider himself having lost a very important component of the person, but it does not automatically mean that the rest of the person no longer exists or is reduced to remnants of one's anatomy, physiology or chemistry. This is not to suggest that a Chinese person would insist on maintaining a (totally) brain dead human body with advanced life support, only to point out that the determination of personhood based exclusively on psychological criteria must be negotiated sensitively with the Chinese conception of the person as a psychosomatic unity. This also suggests that a Chinese would encounter considerable difficulty in accepting futility determinations for one who is in a so-called "persistent vegetative state," for he is likely to be considered no less a person.

5. THE RELATIONAL BASIS OF PERSONHOOD
IN THE CONFUCIAN TRADITION

The emphasis of the physical body and the understanding of the human person as a psychosomatic unity points to the next characterization of the human person as a social and relational being. The early Confucian thinker Hsün Tzu (or Xung Zi) (298-238 BC) recognized that human beings are distinguishable from animals because of their relational/social capacity. He said, "Men are not as strong as oxes, nor do they run as fast as horses, yet how is it that oxes and horses are being mastered by men? That is because men are capable of social organization and animals are not" (*Hsün Tzu*, V: 9, my translation). In the Confucian tradition, a person is never seen as an isolated individual but is always conceived of as a part of a network of relations. Even though it may seem to many that the bulk of Confucian ethical teaching pertains to the self-cultivation of an individual person to become a sage, one must not lose sight of the fact that the whole process is carried out in and through the social context and for the purpose of fulfilling social responsibility rather than self-actualization *per se*. Even though the cultivation of subjectivity as a holistic person was developed and emphasized by later Neo-Confucianists of the Sung Dynasty (960-1279 AD), this so-called subjectivity is located within a social collective subjectivity. Hence, any subjective consciousness that a person has been made aware of through self-cultivation is more a reflection of a collective subjectivity of the community than it is of his own subjectivity. In other words, the Chinese understanding of self-cultivation is never on the establishment of the independence or individuation of the subject, but on the promotion and maintenance of the collective harmony of the community. The entire Confucian program for self-cultivation is to emphasize the social nature of man.

Thus, in the Confucian tradition, to actualize man's nature is to fulfill man's human relatedness; a person is always a "person-in-relations." Mencius said, "Slight is the difference between man and the brutes. The common man loses this distinguishing feature, while the gentleman retains it. Shun understood the way of things and had a keen insight into human relationships. He followed the path of morality" (*Mencius* 4B:19; Lau, p. 165).[1] So the ability to have human relationships is what differentiates human persons from animals, and what is crucial for a human person is to be self-consciously aware of "a keen insight" in human relationships and so "by nature" follow "the path of morality." Mencius was cognizant of the danger that human beings may lose this consciousness of human sociality and relationships and thereby degenerate into behaviors not significantly different from animals. This is why Mencius repeatedly emphasized the importance of education and moral cultivation. When a person is in possession of "a keen insight into human relationships," he becomes a fulfilled and actualized true person, a sage. Mencius said, "The [carpenter's] compass and square produce perfect circles and squares. The sages exhibit perfect human relations" (*Mencius* 4A: 2; Lau, p. 139, my translation). This suggests that in Confucian anthropology, relations are of the ontological category, in that they both constitute and complete personhood. This becomes clear through an analysis of the Confucian notion of *ren (jen)*.

6. THE PICTOGRAPHIC SIGNIFICANCE OF *REN*

The doctrinal core of Confucianism is the principle of *ren*. The word *"ren"* first appeared in pre-Chin classics *The Shoo King or The Book of Historical Documents*, in "The Book of Chow: The Metal-bound coffer": "I have been lovingly obedient (*"ren"*) to my father."[2] In *The Analects*, the term *"ren"* appears 105 times, more than any other term such as heaven, ritual, and filial piety. It has been variously translated as benevolence, humanity, humanness, compassion and so on. Such a wide spectrum of translation reflects the fact that all the other Confucian ideals, humanity, personality, morality, and political ideology, are deduced from this central doctrine. Depending on the different contexts in which the term is being used, it has always been a challenge to try to understand its exact meaning.

To begin with, *ren* is a relational term and is used to describe the relational/social nature of man. The Chinese ideograph for *ren* is composed of two characters, "human" and "two," denoting that whatever else the term may mean, its meaning is intended to be accomplished through human relationships. According to Liu's etymological research (1992, pp. 252-266) on the pictographic meaning of the character, *ren* depicts two simple indigenous people of the Southwest greeting each other. The pictogram reveals the following characteristics: (i) two people are involved, (ii) they are facing each other directly and symmetrically, and (iii) their knees are bent and the upper bodies are in a forward-thrust posture. In other words, the word *ren* is rooted in an ancient ritual/etiquette in which two people are courteously or ceremoniously facing each other, bowing to each other to greet and pay respect to each other. It is a kind of body language which denotes a basic interactive relationship between people, and reveals an interpersonal consciousness marked by mutuality, equality and affectivity found in the ancient tribal people. The bending of knees and the upper torso signifies a benevolent affection; it may also signify a sense of mutual humility. Of note, during the Chow dynasty, the bending of the knees was eliminated, but the bowing movement was retained.

According to the classical interpretation of *Chung-yung* by Zheng, the word *"ren"* means "man" (person), and he explained further that person/man here means "seeing the other the same as oneself and greeting each other to express care". Zheng's exegetical remark captures the importance of the mutual respect implied in the character of *ren* and emphasizes the integrity and dignity of the individual which is often lost in the hierarchical structure of human relations developed subsequently in the Chinese society. But Mencius seemed to have noted this prerequisite for proper human relationships. He says, "Here is a basketful of rice and a bowlful of soup. Getting them will mean life; not getting them will mean death. When these are given with abuse, even a wayfarer would not accept them; when these are given after being trampled upon, even a beggar would not accept them" (*Mencius* 6A: 10; Lau, p. 235). In Mencius' mind, to be truly *ren*, and give charity as a form of *ren*, an act must be accompanied by respect. Individual integrity and dignity are presupposed and not precluded by human relations in which *ren* is actualized.

7. *REN* AS THE DYNAMIC PROCESS OF "PERSON-MAKING"

David Hall and Roger Ames have noted that both *Mencius* and *Chung-yung* have explicitly stated that *"ren* means man", and together with the observation by

Boodberg (Boodberg, 1953, p. 328) that the word *ren* is used in Confucian texts not only as a noun and adjective but also as a verb, these authors argue that *ren* does not only refer to an achieved state of humanity either as some inner character/spiritual condition/virtue, or as an aspect of the outward action/conduct, but as a dynamic process of "person making" (Hall and Ames, 1987, pp. 114-125). They argue that if *ren* as benevolence and *ren** as man both mean "person", the distinction between the two must be qualitative in nature, distinguishing two different degrees of humanity, with *ren* referring to the process by which a "power man" may be transformed to a benevolent person. As *Chung-yung* says, "The realization of oneself is *ren*" (Hughes, 1979). What is crucial to note is that the process of transformation involves an indispensable relational dimension as indicated by the addition of the numeral "two" in the formation of *ren* from *ren**. This means that the benevolent person "is attainable only in a communal context through interpersonal exchange" (Hall and Ames, p. 116). Confucius himself was quite aware of the crucial importance of this interpersonal dimension. His remark to Tzu-lu revealed that clearly. "One cannot associate with birds and beasts. Am I not a member of this human race? Who, then, is there for me to associate with?" (*Analects* 18: 6; Lau, p. 185). Hall and Ames refer to this person-making process as fundamentally an integrative one in which the self "with his disintegrative preoccupations with selfish advantage" is transformed into "the profoundly relational person" (Hall and Ames, p. 115).

The first main feature of Confucius' teaching on this relational person is the ability to show deference to others. For example, when asked by Tzu-chang about benevolence, he replied "There are five things and whoever is capable of putting them into practice in the Empire is certainly 'benevolent.' ... They are respectfulness, tolerance, trustworthiness in word, quickness and generosity" (*Analects* 17: 6; Lau, p. 173). These five attitudes are primarily concerned with showing deference to others and are considered by Confucius to be essential elements in establishing relations. In reply to Fan Ch'ih's question on benevolence, he gave a similar answer, "While at home hold yourself in a respectful attitude; when serving in an official capacity be reverent; when dealing with others give of your best" (*Analects* 13: 19; Lau, p. 127). Confucius recognized that to take relations seriously involves not only an attitude or inner disposition of tolerance, reverence, generosity and so on, but perhaps more importantly, a willingness to conform to the various norms of respectfulness, tolerance, and trustworthiness established by the community. By conforming to the norms or rites (see below), one is showing the highest respect for the community and the standard the community as a whole embraces. On this view, *ren* in Confucius' mind involves not only the willingness of the self to respect others in the community but also the readiness to become part of the community. In this sense, the Confucian person is irreducibly communal, always willing to conform to the demands of the community while not necessarily compromising what he himself believes to be appropriate. *Ren* as person-making means respect for the other and conformity to the consensus of the community. This is quite clearly expressed in a remark made by Confucius when he was asked by Chung-kung about benevolence. The master said, "When abroad behave as though you were receiving an important guest. When employing the services of the common people behave as though you were officiating at an important sacrifice" (*Analects* 12: 2; Lau, p. 109).

In the context of person-making, Confucius also characterized *ren* as loving one's fellow men (*Analects* 12: 22; Lau, p.117). To love in the Confucian sense of actualizing *ren* is to take the other into one's own sphere of concern, interest, value and viewpoint, and in the process, the other becomes an integral part of one's own person. In this exchange each allows oneself to be defined by the other in a mutual and dynamic process of person-making. But from the way *ren* was formulated by Confucius, it is quite clear that the process is not self-effacing but self-referential. *Ren* as love is negatively expressed as "Do not impose on others what yourself do not desire" (*Analects* 12: 2; Lau p. 117) and positively it means that "a benevolent man helps others to take their stand in that he himself wishes to take his stand, and gets others there in that he himself wishes to get there" (*Analects* 6: 30; Lau p.55). This is commonly called *shu*, which can be defined as an assessment or treatment of others using oneself as the analogical point of reference. Confucius considers this ability to use the self as analogy "the method of benevolence" (*Analects* 6: 30). In applying the self as a measure or analogy, one projects oneself into the circumstances of the other to find out what the other person wants or does not want; but this does not entail that one abandons one's own identity or compromises one's own values. Rather, "One acts by extending the perimeter and parameters of one's own person to embrace the defining conditions, perceived attitudes, and the background of the other person in order to effectively become "two" perspectives grounded in one judgment" (Hall and Ames, p. 117).

These two features of person-making, deference to others and retaining the self as the point of reference, suggest that there is a mutuality involved in the person-making process of *ren* in that as one influences the other, one is simultaneously under the influence of the other. Hall and Ames explain the interdependency and interpenetration in the process of person-making like this:

> The importance and influence of a person becomes measurable in terms of the extension into and integration with the selves of others. That is, a person is meaningful and valuable as a function of his participation in the field of selves that constitutes his community, and the quality of his own person in turn is a function of both the richness and diversity of the contributing selves that he has brought into his particular focus, and the extent that he has been successful in maximizing their creative possibilities (Hall and Ames, p. 119).

Ren as a process of person-making through love therefore is meant to be a reciprocal process, so that in mutually depending on each other and penetrating each other, Confucius expects a bonding of relationship due to mutual reciprocation. In the actualization of *ren*, on the one hand, there is the extension of oneself outward to the other; on the other hand, there is the appropriation of the other inward to the self; in the midst of this exchange of each other, there is the resultant reciprocal bonding of one and other.

Since Confucius believed that the person-making of *ren* involves the mutual incorporation between the self and the other, there is a strong sense that the person-making development of both the self and the other is coextensive and coterminous. This is so because if person-making involves a mutual appropriation of each other, it follows that there is a close correlation between the cultivation of personhood or *ren* of both the self and the other. In a passage recorded by Hsün Tzu, Confucius made three different but related comments on *ren*: (i) "a benevolent person is one who causes others to love him"; (ii) "a benevolent person is one who loves others"; and (iii) "a benevolent person is one who loves himself" (*Hsün Tzu* 20: 29). As the person-making process progresses, there is an increasing mutual incorporation between the self and the other so that eventually the self is so united with the other within the matrix of the united "self-other" interest that to love self and the other is no longer distinguishable. Simultaneously the self is constituted in the process to become a person and Confucius considered *ren* as self-love the highest form of moral character. In this sense, the focus of Confucius' person-making (*ren*) is not merely on the relational self, but equally importantly on the relational matrix within which the person eventually emerges. The Confucian person is constituted by the relation that exists between the self and the other, both of whom attain personhood in that matrix. In this regard the Confucian person bears certain resemblance to the Christian perichoretic personhood in being on onto-relational concept.

This aspect of reciprocation and bonding as well as the emphasis on the importance of the relational matrix in person-making are some of the main reasons why Confucius taught that the love of *ren* should begin with one's family members first before one extends the process to the society at large. Confucius made this plain' when be said that "Being good as a son and obedient as a young man is, perhaps, the root of a man's character" (*Analects* 1: 2; Lau; p. 3). Clearly Confucius believed that if a man cannot even reciprocate the intimate human affections experienced within the family bond, it is not likely that person-making can be achieved in the less intimate environment outside the family circle. This is seen in Confucius' teaching of "Three year mourning" which required children to keep a three year mourning period upon the death of their parent. Even at Confucius' time, people were challenging the rule, and one of Confucius' own students Tsai Wo was also critical of the requirement, complaining that the rule was disruptive and impractical. Confucius' judgment on the issue was that the abandonment of the "Three year mourning" is a manifestation of the lack of benevolence (*ren*) (*Analects* 17: 21). It is significant to note that Confucius did not consider such an abandonment a form of filial impiety, which seems to be a more logical thing to say; instead, he called it a form of non-benevolence (not in keeping with *ren*). From the reason Confucius gave for his judgment, we may gain a glimpse of the emotional and affective content of *ren* which Confucius had in mind, as well as the notion of reciprocity inherent in the person-making process of *ren*. Confucius said: "A child ceases to be nursed by his parents only when he is three years old. Three year's mourning is observed throughout the Empire. Was Yü not given three years' love by his parents?" (*Analects* 17: 21; D.C. Lau, p. 179). In Confucius' mind, the basis of the "Three-year mourning" requirement for one's deceased parent is based on the bond of love, brought about by the parents' initial loving relationship to the child. Confucius expected that such bonding of love should become part of any person's nature. Hence for Confucius, the requirement is therefore not a legalistic accounting between a later generation to a former one; rather it should be a spontaneous expression of a person's nature in reciprocation to an earlier expression of love of the person's deceased parent. To be willing to give up the "Three-year mourning" is therefore much more than a failure to perform certain filial duties or rites. It shows that one is deficient in the love-bonding of the person; hence it is called "non-benevolent." In other words, if a man fails to reciprocate his parents' love as expressed by his refusal to keep the "Three year mourning," the person-making process of *ren* has failed and he has not attained personhood.

8. CHINESE PERSONHOOD AND BEGINNING OF LIFE ISSUES

In the Chinese conception of social personhood, the definition of a person as a "Being-in-family relations" has significant implications for the beginning of life ethical issues. For example, in the relationship between father and son in the family, one of the most important ways to show filial piety is to provide male offspring to propagate the family name. Mencius has specifically stated that, "There are three things which are unfilial, and to have no posterity is the greatest of them all" (*Mencius* 4A: 26; Lau, p. 75). This explains why births in general and male births in particular are welcome events in Chinese society. Therefore, traditional Chinese attitude towards abortion is generally negative, and this has also been reinforced by the traditional teaching of the unity of nature in which humanity is not only a member of the Heaven-Earth-Man triad, but also a product of Heaven and Earth. Life is always viewed as precious, and the taking of life is something to be approached with the utmost caution. This high view of life is supplemented with the Buddhist teaching of compassion which was imported to China from India around the second century AD.

Since the value of male progeny as a means to continue the family name assumes an especially important place in Chinese culture, it will be consistent with traditional Confucian thinking for a Chinese to endorse the use of modern assisted reproductive technologies to ensure male progeny as long as the sperm comes from the husband. The use of donor eggs and the employment of surrogate mothers would probably be less desirable, but they are not of primary importance. This tradition of producing a male offspring as fulfilment of filial piety may also be seen as a strong justificatory reason to perform prenatal sex selection in contemporary China where for reasons of population control the policy of "one child per family"[3] is being practiced.

9. CHINESE PERSONHOOD AND DEATH AND DYING

Confucius himself accepted death as an objective fact of the end of life. He was known to have made terse, factual and matter-of-fact pronouncements about his students' and friends' death without much explanation, as if there was nothing to be explained. To him death is regrettable and unavoidable, but death is not the most significant event in life. As *ren* (humanity, benevolence) is the most important "life-guiding" principle, it also regulates the process of dying as well as the significance of death. Confucius said, "A resolute scholar and a man of humanity will never seek to live at the expense of injuring humanity. He would rather sacrifice his life in order to realize humanity" (*Analects* 15: 8; Chan, p. 43). In other words, if *ren* is achieved and human personhood is fulfilled, then death pales in significance. "The Master said, 'He who has heard about *tao* in the morning, may readily die in the evening'" (*Analects* 15: 8; Chan, p. 43). For a Chinese therefore, death is evaluated in terms of one's fulfilment of *ren*. It is a "good" death, a worthy and acceptable one, only when most, if not all, of one's moral duties of life have been properly fulfilled (one has heard about *tao*). Such an ethic of relational and moral personhood is likely to impact the attitude and course of action of a dying patient in two different ways. On the one hand, any resistance to acknowledge a certain terminal illness or to forego futile medical treatments may mean some unfinished business perceived by the patient, and hence the effort to extend life to the last breath in order to complete one's unfinished task(s) in life. On the other hand, one may readily give up treatment and courageously encounter death to avoid becoming a burden to one's family or community as a way of fulfilling *ren*.

Furthermore, the requirement of filial piety as part of the social personhood also makes a difference in the acknowledgment of the terminal nature of an illness. For example, an elderly dying Chinese person may conceivably resign herself to a "good death," but her children are reluctant to grant her this wish for "good death" for reasons related to filial piety. Since filial piety can only be expressed when a parent is alive, to extend an ailing parent's life is to extend the opportunity to show filial piety. This alone is a common reason for children to reject a physician's judgment that any further medical intervention is futile, and to insist that every heroic measure be done for their dying parent. To forfeit this duty produces a strong sense of guilt. Sometimes, the sense of guilt due to failure to show sufficient filial piety to a parent or family elder prior to the elder's terminal illness can be a significant additional reason to request unreasonable life-extending measures. Another strong expression of filial piety is found in the saying that, "The body, including our hair and skin, we receive from our parents. We dare not cause any injury to it, and this is the beginning of filial piety."[4] This thought is so deeply ingrained in the Chinese mindset that it is a decisive reason why one should take care of one's health in all circumstances. The notion that physical harm to one's body amounts to a violation of filial piety also accounts for many instances of persistence in medical treatment even when such treatment has been determined to be of no further medical benefit.

10. CHINESE PERSONHOOD AND INFORMED CONSENT

The Chinese concept of holistic and social personhood, regulated by the principles of *ren* and filial piety and organized in a social hierarchy, provides distinctive perspectives when issues of informed consent in health care ethics are being considered, particularly in the case of an elderly patient. From the physician's point of view, both his role and function are impacted by the Chinese understanding of personhood. To start with, Chinese believe that the principle of *ren* should be the regulatory principle for the physician's character as well as his conduct just as it is for other members of the community. For this reason, a Chinese physician will likely follow the inclinations of the family to satisfy all the other social protocols inspired by the principle of *ren*, particularly the requirements of filial piety as perceived by the patient's family. On the other hand, even though the exact position of a physician is not specified in the hierarchical organization of a Chinese society, traditionally physicians command a respect equivalent to that of parents in family or elders in a clan. There is a common saying in Chinese: "Physicians have the heart of parents," indicating that people expect physicians to act lovingly as parents do, and accord them a level of respect commensurate with elders and parents. This means that the physician is in a social position to override the decision of the family if he feels the need to do so. With this venerable status, it is not uncommon for Chinese physicians to have a paternalistic attitude in their dealings with patients. They will not find it ethically troubling to withhold information from patients if in their judgment it is not beneficial for them. As well, in keeping with the Confucian tradition, patients and their family find it quite natural to submit to the opinion of their physician. In this regard, the inclinations of patient's family and physician tend to act synergistically to disrespect a patient's autonomy.

From the patient's point of view, the imperative to respect an individual's right to self-determination, considered to be sacrosanct in the West, is simply not required by the Chinese ethic of personhood. In fact, the Chinese concept of a more holistic and social personhood challenges the assumption that the patient is the one to be told of the diagnosis and prognosis and to make medical decisions. When a Chinese becomes sick, his behavior and expectation are closely tied with a strong sense of personal identity with and mutual dependence on his family. For example, it is not shameful for a Chinese elderly sick person to be dependent on her children, but rather it is something to which she is entitled. According to *the Book of Rites*, the rules are: "When a ruler is ill, and has to drink medicine, the minister first tastes it. The same is the rule for a son and an ailing parent" (*The Li Ki*, Bk 1, Part III, sec. 2, 1966, Legge (trans.), p. 114). In the Chinese social hierarchy, the elderly sick person can expect to be cared for by his family, and his sick role includes the privilege to be relieved of a large share of personal responsibility, including most of the decision-making process of her own medical care, even though the patient may be rational and competent. Family members, especially the patient's children, are expected to take over that responsibility and assume the various roles of being child, caregiver, protector, and surrogate decision-maker. Whereas in the West, family inputs are usually not determinative, in the Chinese society and for that matter in any Asian culture where Confucianism exerts considerable influence, family inputs are often decisive, making the conventional requirement of patient informed consent not easily enforceable.

If the Chinese ethic of personhood generally excuses elderly patients from being informed of their medical condition and sharing in the responsibility of their own medical care, it specifically shields patients from being told the news of a terminal illness, and/or the judgment to withhold or withdraw futile treatments. Whereas in the West good bioethical practice would consider it both appropriate and necessary to provide the patient with all the medical facts, including the fateful ones, directly, fully and truthfully, a Chinese ethic of personhood, expressed as the rule of filial piety, may consider it morally inexcusable to disclose terminal illness which may add further harm to the patient. To fulfill filial piety, dying elderly patients must be protected from fateful news, and this of course is at odds with the Western notion of informed consent.

11. CONCLUSION

The distinctive Chinese understanding of personhood has much to contribute to modify the narrowly psychological and individualistic understanding of personhood and the rights-based ethic derived from it currently prevalent in the West. For example, in most Western countries, it is entirely a matter of a woman's right or autonomous choice to abort a fetus. The Chinese holistic and social personhood suggests that the personhood of the fetus could very well have been established by the materno-fetal relation. Such an ethic may also suggest that the sexual partner's opinions and desires may not be set aside entirely as is done in the West. The Chinese ethic of a relational personhood also challenges many questionable practices in the area of assisted reproductive technology. Specifically, the anonymous donation of human gametes, gestational surrogacy arrangements and the possibility of asexual reproduction are examples in which the relational context not only of human personhood but also of human existence has been totally ignored. A relationally defined personhood may also modify the way certain patients are treated, specifically those with severe mental retardation, in irreversible coma or persistent vegetative state. The different understandings of personhood in the West and the East emphasize the importance of trans-cultural ethical discourse. For better or for worse, we live in a world which has become squeezed into a global village. Such a dialogue becomes necessary if confrontation and conflict are to be avoided. For a human being to be a person, both individuality and relationality are constitutive components. In this context, the East and the West have a lot to learn from each other and to benefit each other.

Regent College
Vancouver, British Columbia, Canada

NOTES

[1] Shun is the legendary ancient model sageking believed to have lived around the 3rd millennium BC and who successfully built an ideal society.

[2] See James Legge (1985, p. 354). Legge's footnote states that Medhurst translates the clause as "my benevolence is equal to that of my forefathers."

[3] Put more precisely, China's policy of population control is "one, two, three, four and no limit." That is, for Han people (more than 90% of Chinese are Han) living in cities, one family can only have one child; in rural areas they can have two; for other minorities one family can have three or four children; and for Tibetans, there is no limit for their family.

[4] See *The Book on Filial Piety*, my translation.

REFERENCES

The Analects, 2nd edition (1992). D. C. Lau (Trans.). Hong Kong: The Chinese University Press.

Boodberg, Peter (1953). 'The semasiology of some primary Confucian concepts,' *Philosophy East and West*, 2, 317-332.

Chan, W.T. (1963). *Source Book in Chinese Philosophy*. Princeton: Princeton University Press.

Dennett, D.C. (1976). 'Conditions of personhood.' In: A. O. Rorty (Ed.), *Identities of Persons*. Berkeley: University of California Press.

Descartes, R. (1995). 'Discourse on method, Parts 4 and 5.' In: Hunter Brown, Dennis L. Hudecki, Leonard A. Kennedy & John J. Snyder (Eds.), *Images of the Human* (p. 155). The Philosophy of the Human Person in a Religious Context. Chicago: Loyola Press.

Engelhardt, H.T., Jr. (1996). *The Foundation of Bioethics*, 2nd edition. New York: Oxford University Press.

Fletcher, J. (1975). 'Four indicators of humanhood – The enquiry matures,' *The Hastings Center Report*, 5 (December), 4-7.

Frankfurt, H. G. (1971). 'Freedom of the will and the concept of a person,' *Journal of Philosophy*, 68(1), 15-20.

Hall, D.L. & Ames, R.T. (1987). *Thinking Through Confucius*. Albany: State University of New York.

Hartshorne, C. (1981). 'Concerning abortion: An attempt at a rational view,' *The Christian Century*, 98(January), 42-45.

Hsun-Tzu (1970). *Hsun-Tzu (340-245 B.C.)*.

Huang Ti (1970). *Huang Ti Nei Ching Su Wen (The Yellow Emperor's Classic of Internal Medicine)*, Ilza Veith (Trans.). Berkeley: University of California Press.

Hughes, Ernest Richard (1979). *The Great Learning & the Mean-in-action*. New York: AMS Press.

Legge, J. (1985). *The Chinese Classics*, Vol. III, *The Shoo King*. Taipei: Southern Materials Center, Inc.

The Li Ki (1966). In: James Legge (Trans.), F. Max Müler (Ed.), *The Sacred Book of the East*, Vol. XXVII. Delhi: Motilal Banarsidass.

Liu, W. Y. (1992). 'The original meaning of "ren" and its two basic specifications.' In: *The Proceedings of a Conference in Memory of the 2540 Birthday of Confucius* (in Chinese). Shanghai: San Lian Book House.

Mencius (1984). *Mencius*. D.C. Lau (Trans.). Hong Kong: The Chinese University Press.

Tooley, M. (1973). 'A defense of abortion and infanticide.' In: Joel Feinberg (Ed.), *The Problem of Abortion*. Belmont: Wordsworth.

Traylor, Kenneth L. (1988). *Chinese Filial Piety*. Bloomington, IN: Eastern Press.

Walters, J.W. (1997). *What Is a Person? An Ethical Exploration*. Urbana: University of Illinois Press.

CHAPTER 4

HYAKUDAI SAKAMOTO

THE FOUNDATIONS OF A POSSIBLE ASIAN BIOETHICS

1. INTRODUCTION

My Asian colleagues and I understand that, in Asia, studies in bioethics stand far behind in terms of the Western standard. However, bioethics in the Asian region may be fundamentally different from bioethics in the West. The Asian bioethical pattern should have its own cultural, ethnological and philosophical basis, reflecting the present multi-cultural post-modernism, which largely differs from the Western circumstance. In recent years, I have expressed this view at several international conferences and congresses, including the Fukui Seminar on Bioethics in Japan in 1991 and the 1993 2nd World Congress of the International Association for Bioethics in Buenos Aires, Argentina. In some cases my view was sympathetically welcomed, while in the others it was strongly rejected from the universalists' point of view. In the following, I will discuss my philosophical ideas on the foundations of a possible Asian bioethics.

2. WHAT IS THE NATURE OF WESTERN BIOETHICS?

To begin, let me examine the essential nature of Western bioethics. In the Western world, bioethics, distinguished from traditional medical ethics, emerged in the late 1960s as a result of the innovation of science and technology and the movement of Technology Assessment (TA) promoted in advanced nations, especially in the U.S.A. The TA movement was also introduced in Japan. The Japanese government and society welcomed the TA movement, fearful of unexpected hazards caused by new, large-scale technology innovation. The TA movement also came to Korea, but not to China because of its social and political situation.

Now, what was the Western criterion of assessing science and technology during the 1960s and 1970s? I assume it was "humanism". Some sciences and technologies were banned because they were anti-humanistic. However, what is humanism? Historically speaking, humanism in the Western world was essentially "human-centricism", which

R.-Z. Qiu (Ed.), Bioethics: Asian Perspectives, 45-48.
© *2003 Kluwer Academic Publishers. Printed in the Netherlands.*

was backed up by the human "frontier mentality" according to the terminology of Daniel Chiras (1985). Also, this humanism was fortified by the idea of "person" and "human dignity" significantly emphasized in the 18th century philosophy of, for example, Immanuel Kant. Thus, "person" is identified and dignified as a rational being, and therefore a human being, as a person, is given human rights, especially fundamental human rights such as the heroic freedom to conquer Nature.

At the first stage of the Technology Assessment, the criterion of the assessment was clearly to protect human beings from any technology disaster, and this aim was easily identified with protection of human rights. This general mood reflected on the bioethics of the first stage, which describes "bioethics" as "the way to protect human rights from biotechnologies" and through the course of debates on this issue, the traditional paternalism was generally categorically rejected. From this point of view, "self-determination", "informed consent", "patients' rights", etc., were recommended, and paternalistic attitudes (usually of physicians) were criticized. In this stage, almost all issues of bioethics were treated by a principle of "protection of human rights." For instance, in the U.S.A, "bioethics" meant the "establishment of a legal system addressing bioethical issues from the viewpoint of human rights," as Alexander Capron suggested to me in Buenos Aires.

3. A TURNING POINT OF BIOETHICS IN THE 1980S

However, bioethics came to a turning point in the 1980s. This was brought about by, first, the extremely rapid development of genetics; second, by the rise of environmental approaches to bioethics; and third, by the inclusion in bioethics of a consideration of Asian and other non-European paradigms. First, by the end of the 20th century, we had almost obtained the ability to manipulate the human genome, by way of recombinant DNA (i.e., the ability to alter the genetic character of human body artificially). This implies the possibility of "artificial evolution." But how is this justified bioethically?

The first apparent bioethical attack on genetic engineering was made by the Council of Europe of EC in 1982 by its Recommendation 934 on genetic engineering from the standpoint of human rights. It says: "Human rights imply the right to inherit a genetic pattern which has not been artificially changed" (Annuaire Europeen-European Yearbook Vol. 30, 1983). However, we are now going to admit "gene therapy" which necessarily changes the human genetic pattern artificially by the name of "medical treatment", which promises "human happiness." Here, human happiness has possibly become a concept which contradicts the protection of human rights.

Second, conflicts occurred between two different types of concepts of "protection of the environment." One takes the point of protecting the environment to be preserving the best living conditions for present or future generations of humans. The other takes the point to be protecting Nature for its own sake. The former is typically human-centric, and it has gradually been replaced by the latter under the influence of recent developments of ecological knowledge together with the severe regret and criticism of the "frontier mentality" of modern humanism. People now tacitly accept the value of

Nature itself, instead of its value for human beings, as the reason for defending the environment.

Third, the range of vision in bioethics has been expanded to regions outside Europe and America, especially to Asia. Many bioethical incidents happened in Asia which were quite strange from the European and American perspectives. For instance, the Japanese rejection of heart transplantation from the brain-dead body for almost thirty years was quite odd for Euro-American minds. The "one child policy" of China is also strange from a European and American point of view. Also, the Malaysian hostility to human rights policy is difficult to understand from a Euro-American perspective. People have begun to notice the peculiarity of Asian minds in considering bioethical issues. Something is fundamentally different.

First of all, in many countries in East and Southeast Asia, the sense of "human rights" is very weak, and they have no theoretical background for the concept of human rights. Rather, they are concerned with overcoming starvation and poverty not by human rights, but by mutual aid and new technology. The recent introduction of the European idea of "human rights" has caused ethical and moral conflicts. The Asian view of Nature is also historically heterogeneous from the European vision. Nature is something not to be conquered but to live with. Generally speaking, the Asians have a holistic way of thinking which is distinct from the European individualistic way of thinking. Therefore, Asian people put higher value on holistic happiness and the welfare of the whole group or nation to which they belong rather than on their individual human rights.

In the present post-modern age, it may be necessary for human society to globalize bioethics for its future development. But it is impossible to do this by insisting on the universality of human rights, namely, the universality of the Euro-American standard of bioethics. This is the reason why the articulation of a possible proper Asian bioethics is now necessary.

4. THE NATURE OF THE ASIAN BIOETHICS

First, the new Asian bioethics should stand on the new philosophy concerning the relation between Nature and human beings. At least, the "frontier mentality" of human-centricism of the 18th-century-type humanism must be abandoned. Also, the simple-minded naturalism of "laissez faire" is impossible, for we already have acquired the ability and technology to control human future and human evolution. We should now establish a new humanism without human-centricism, and also cultivate a new methodology to compromise this new humanism and the modern science and technologies to control human evolution.

Secondly, we should reconsider the nature of human beings apart from the theory of the 18th-century philosophical anthropology of Immanuel Kant and other Idealists, which gave a ground for the idea of the universality of human rights. Instead, the equality of non-human rights (i.e., the rights of non-human beings) should be considered. At the same time, the idea of the dignity of the human being should be reconsidered. Why is the human being exclusively dignified? In some traditional ways of thinking in Asia, there is no idea of human dignity distinguished from animal dignity. We must take the standpoint of "value relativism," which is a challenge to ongoing Western bioethics.

At the same time, we have to appreciate the fact that in Western societies, most people, even professional ethicists, are still inclined to believe in the absolute universality of "human rights", and this idea properly functions in leading and controlling their social systems, especially the legal systems in the Western world. Our urgent task here is to find a way of making both positions concerning "human rights" compatible in order to find a new refined methodology of global bioethics.

Thirdly, we have to investigate the new philosophy for the foundations of Asian bioethics. I believe it must be grounded on the Asian ethos, which is fundamentally different from the European ethos in many aspects, and has experienced a different history from the European one, especially in the last few centuries. What, then, is the "Asian ethos"? The Asian ethos is many-sided. Its most remarkable characteristic is its holism in contrast to European "individualism." Taoism, Confucianism, and Buddhism still deeply influence the Asian ethos. Their doctrines and precepts are all holistic in general. They tend to put higher values on Nature, society, community, neighborhood and mutual aid than individual ego human rights. It is a sort of severe anti-egoism. But it is not necessarily altruism either. It always seeks some sort of holistic harmony of antagonists. One might be afraid that this kind of holism is a sort of paternalism, which was rejected in the early days of bioethics in favor of personal autonomy. However, we should notice that some new bioethical issues, such as issues of genetics and environmental crises, necessarily require some sort of paternalism, not of a Western type, but rather of an Asian type. Here, "harmony" is a key word. The Asian bioethics will begin now with the effort not only to deny the European idea of individual autonomy, but also to harmonize it with the new holistic paternalism of our own Asian traditional ethos.

Department of Philosophy
Nihon University
Tokyo, Japan

REFERENCES

Annuaire Europeen-European Yearbook Vol. 30 (1983). Hague: Martinus Nijhoff Publishers.
Chiras, D. (1985). 'Environmental ethics: Foundations of sustainable society.' In: D. Chiras (Ed.), *Environmental Science: A Framework for Decision Making*. San Francisco: Benjamin/Cummings Publishing Company, Inc.

PO-KEUNG IP

CONFUCIAN PERSONHOOD AND BIOETHICS

1. INTRODUCTION

Though it is said that Confucianism endorses a relatively rich notion of a moral person, how that notion of moral personhood relates to bioethics has seldom been clearly articulated and critically scrutinized.[1] It is argued that the familial collectivism entailed by Confucianism is either incompatible with, or too weak to sustain, a rights-based notion of personhood. The Confucian moral architecture, as represented by the *ren-yi-li* normative system, largely underdetermines a notion of a rights-based personhood which is vital to bioethics. Institutionally, the family, society and the state as basic institutions provide the major institutional factors which function as external (social) inhibitors for the development of a rights-based person. The meshing together of these conceptual and institutional factors has constituted a fatefully inhibitory barrier to the development of a rights-based person. This paper examines the bioethical ramifications of the Confucian non-rights-based notion of personhood in areas including abortion, treatment of newborns and doctor-patient relationships.

2. THE SIGNIFICANCE OF PERSONHOOD

Let us begin our discussion with a case in Bioethics:

> A pregnant woman was found by ultrasound examination to be carrying two fetuses. Unfortunately, one fetus did not move and was believed to have died in uterus. The pregnancy was continued until thirty weeks at which time a Cesarean section delivery was performed. Twin *A* was born with no problems other than those routinely associated with prematurity. Twin *B* was born alive, not dead as was suspected from the ultrasound exam. However, it is difficult to be certain about what to call this second product of conception. Twin *B* was alive but did not bear any resemblance to a human fetus. There were no limbs, head, or differentiated internal organs. The mass of tissue was alive, profused by an umbilical cord that was joined to the umbilicus of Twin *A*. Obviously, when the umbilical cord was severed the tissue died. How should this mass of tissue be described? Did it have a right to live? Were the physicians obligated to attempt to sustain its life? Were birth and death certificates necessary? (Shelp 1986, pp. 108-109).

In the case described, whether we have obligations to the newborn depends very much on whether we regard it as a person. To regard it as person is to put it within

R.-Z. Qiu (Ed.), Bioethics: Asian Perspectives, 49-56.
© *2003 Kluwer Academic Publishers. Printed in the Netherlands.*

the moral community, which entitles it to be treated on a par with other members of the community. This demands that members of the moral community should discharge duties and obligations to the newborn. By the same token, being respected as a person can entitle the newborn to put rightful claims upon us. Beyond the treatment of newborns, how we treat the terminally ill, make decisions on organ transplant, reproduction, euthanasia, abortion and patient care also depend on the status of the patient. The determination of personhood in effect provides a standard according to which decisions to provide or withdraw care, to rescue or withhold life-saving support, among others, are made.

But the meanings of personhood are by no means straightforward and unambiguous. They are also reflections of the values and commitment of cultures, communities and individuals. The debate over the meanings of personhood has taken place in the context of abortion. But the significance of the notion is by no means confined to the area of abortion. Other areas, including artificial human reproduction, euthanasia, and organ transplant also critically involve this notion.

The concept of personhood is both descriptive and normative. It is descriptive in that it depicts certain properties or qualities a certain entity possesses. The possession of certain properties itself qualifies an entity to be a person. But the concept also has a normative function. Being qualified as a person requires that certain duties and obligations may rightfully be demanded of us.

Two representative notions of personhood currently under debate are genetic personhood and property-based personhood (Shelp, 1986). The genetic notion of personhood defines a person as an entity having the genetic make-up of the human species. In other words, being a member of human species is the defining characteristic of a person. The property notion of personhood, on the other hand, uses non-genetic properties to define personhood. It views a person as an entity having certain properties or qualities which enable it to be a subject of rights and duties, to be a being with its own ends.

3. MORAL PERSONHOOD IN CONFUCIAN MORAL ARCHITECTURE

How does the Confucian person fit into the picture? It seems that the notion of personhood as conceived within Confucianism is basically articulated in moral terms. The genetic notion seems alien to the Confucian framework. But does Confucianism endorse some kind of property notion?

It seems that Confucianism would subscribe to some sort of property notion, with properties strictly couched in moral terms. But the Confucian notion should be sharply distinguished from other similar notions because it is basically a non-rights-based notion of personhood. For the Confucian, persons may not be bearers of rights. However, in order to understand the Confucian notion of personhood, we must understand familial collectivism, which is at the core of Confucianism.

Confucianism is basically a *humanistic* and *collectivistic* moral enterprise. Since the family sits at its core, it may aptly be deemed *familial collectivism*. The moral enterprise of Confucianism is built on the *ren-yi-li* normative structure. The structure represents, as it were, the overarching moral architecture which defines and sustains morally and socially acceptable conduct and attitudes. *Ren* refers to both a capacity and act of utmost benevolence and love. It is both a capacity to love humanity and other living things as well as an act of doing so. The exercise of *ren* generates a host of benevolent acts and conduct (the manifestation of *ren* in moral practices) which collectively or singly represent a myriad of morally desirable conduct, or behavior exemplars. These behavior exemplars themselves also help to constitute a web of , desirable and acceptable moral norms for personal and social behaviors. Buttressing *ren* is the concept of *yi*, which means moral rightness and appropriateness. It is also a moral norm for conduct and decisions. As regards *li*, it represents the etiquettes, norms, mores and protocols in both everyday and institutional lives. The basis for judging whether a norm is morally acceptable has to rely on *yi* and *ren*. They are, as it were, the final yardstick for moral correctness. Indeed, it is in effect *ren* and *yi* which are capable of dictating or legitimizing the moral acceptability of a certain behavior and conduct. Broadly understood, the *ren-yi-li* normative structure, with its generated array of morally acceptable conduct, indeed provides an elaborate set of norms and moral directives governing and dictating conduct and attitudes in different aspects of an individual's personal as well as interpersonal life. These norms and directives are morally and socially binding.

Confucianism is a virtue-based morality. Within its moral architecture, there is a system of virtues for guiding behavior and conduct. Typical among these major virtues are: trustworthiness, loyalty, courage, respect, filial piety, friendliness, compassion, righteousness, and reciprocity. Confucianism entails a deontological ethics. Except for some minor aberrations, it embraces a non-utilitarian moral philosophy. The morality of an action is to be judged on some moral precepts independently of the consequence of that action.

In traditional China, central to the set of Confucian norms was *filial piety*. As a cardinal principle, it dictated the desirable and proper human relationships within the family and beyond. The extension of this principle in the political sphere generated the principle of loyalty. Also, it had a central place in the moral architecture: Filial piety is fundamental to *ren*.

As a result of the paramount importance given to filial piety, it dominated and defined all other salient human relationships. On the other hand, if filial piety was regarded as the cardinal principle governing relationships within the family and the state, it underscored the *hierarchical* structure of human relationships within the family as well as the state, as filial piety itself demands a hierarchical human relationship. To the extent that filial piety centrally endorses hierarchical human relationships, it runs contrary to the value of equality of persons that sits at the heart of the rights-based concept of a person.

4. MORAL PERSONALISM

To fully appreciate the Confucian conception of moral personhood, we have to turn to its conception of moral development (referred to here as moral personalism) as best presented in the *Great Learning*. There, how one can develop one's moral personhood is explicitly articulated.[2] Central to the Confucian moral psychology is the "innate goodness" assumption, which in effect states that humans by nature are good. Evil is only a result of external factors, including society.[3]

According to moral personalism, to achieve moral self-cultivation is a person's most fundamental endeavor. Only by achieving this goal can the person be able to take care of a family, manage the state, and govern the world. The question is whether this Confucian moral personalism, together with familial collectivism, suffice to build up a robust rights-based concept of personhood. Under moral personalism, moral self-cultivation could be interpreted as a development process from a person's personal moral self (person) to his interpersonal moral self (person), understood here as a rights-based person. But the personal moral self should be accomplished *prior* to achieving the interpersonal moral self. The development of the rights-based self begins only after the personal self completes its development. However, the development of the full moral person may not follow a natural progressive pathway, but rather may progress forward in leaps and bounds, or may fail to move forward at all. A person's personal moral development may have little to do with his civic development. In other words, a successfully developed personal moral self is no guarantee for the emergence of an equally successfully civic self.

There are factors governing the developmental transition from the personal to civic self. The absence of these factors can effectively arrest the development of a full moral person, hampering his progress into the stage of a civic person. The moral person thus remains a non-rights-based moral person. The Confucian moral person historically seems to be restricted to a stage of retardation in moral development. Institutional factors put their toll on the moral person.

5. THE FIVE RELATIONSHIPS AND THE FAMILY

The social and institutional manifestation of the Confucian collectivism is the family or the clan. Throughout Chinese history, the most prominent basic institution of China has been the family.[4] The state in the form of dynastic imperial kingdoms was in effect nothing more than the political extension or amplification of the family, albeit one victorious family in a cyclical power struggle for domination. As a result of the Confucian collectivistic leanings, the impact of the family on the individual in the traditional Chinese context was profound. How human relationships were conceived and structured within the Chinese family helped explain why the individual and his rights failed to emerge prominently and flourish in traditional as well as modern Chinese culture. Within the family, *hierarchical and paternalistic relationships* dominated. The hierarchical and paternalistic social structure was successful in preempting equals among persons within the family. The notion of *equal persons* and *equal respect for persons* simply did not have a place here. Permissible conduct or behaviors within such a social setting would take the form of *obligations* and *duties* rather than rights. This was coherent with the Confucian collectivistic ethics which put the value of the collectivity, i.e., the family, above that of an individual.

On a broader scale, there were the *five basic Confucian human relationships*: emperor-officials, father-son, brother-brother, husband-wife and between friends. With the exception of friends, they presuppose a structure of subservience and domination. Again, equality and equal respect simply had no place here. In effect, these salient human relationships in society were nothing more than different manifestations of the familial relationships. There were three cardinal norms which governed and constrained the five salient social relationships. They were highly status-bound and status-regarding norms. The emperor, the father, and the husband, for example, occupied these norm-generating and norm-dictating statuses and positions. They invariably demanded a subservience and respect from the official, the son and the wife respectively. There was no reciprocal respect necessary from the former to the latter. Indeed, respect was non-reciprocal and one-way. As a result, the relationship was again one of domination-*cum*-subordination. Other social relationships and configurations would spin around these unequal and domination-*cum*-subordination social arrangements. To reinforce this domination-subordination relationship was the norm of filial piety as alluded to earlier. There were mutual reinforcements between ideology and institutions in traditional China. Confucian ethics provided the moral and ideological foundation for the family. The development of the family as an institution through its practices in turn helped to sustain the correctness of Confucian ethics which lay at its very foundation.

To the extent that Confucian familial collectivism is inhibitory to rights and equality, it is conceptually incompatible with the concept of a rights-based personhood. One may argue that Confucianism in its formal version may underdetermine the idea of a rights-based person. Or one may argue that within the Confucian moral architecture of *ren-yi-li*, there is no explicit rejection of rights, nor is there a strong endorsement of all forms of inequality, even though the preference for filial piety does entail some forms of social inequality. However, it is familial collectivism which is the core of Confucianism, and substantive Confucianism contributed critically to the underdevelopment of the rights-based person. On the basis of the above deliberations, it seems that a non-rights-based concept of personhood is more in line with the philosophy of Confucianism.

6. CONFUCIAN PERSON IN BIOETHICS

Personhood is not only important in deciding the moral status of fetuses, but also provides the philosophical underpinnings for agents as givers and/or receivers in the bioethical context. Different notions of personhood may generate for the agents as persons different arguments, views, concerns, and courses of actions.

A non-rights-based Confucian person can be a virtuous person. But he may not be rights-regarding. What would be the implications of such a notion in situations like abortion, euthanasia, organ transplant, human experimentation, and artificial human procreation? In particular, what are the implications of such a notion when the rights and equality of persons are critically involved?

How would a virtuous non-rights-based person react to these situations? What decisions would be made by him? On what basis? What policies would be recommended? What actions would be taken? Can the *ren-yi-li* normative structure provide him with the proper guidance? Can Confucian benevolence enable him to make the right choices? Can the moral framework he would be subscribing to provide him good advice to resolve the moral dilemmas, to make a balanced response? It seems that the situations involved are so complex that they defy a blanket analysis. Each situation is complex enough to warrant separate deliberation.

Let us refer to the case presented at the start of the discussion. How would a Confucian person respond to Twin *B*? As a physician, being virtuous and benevolent (though not a rights-regarding person), he would try his very best to sustain its life. He may or may not subscribe to some genetic notion of personhood. If he does, he surely will try to save Twin *B*. If he does not, he will also likely try to save it as he would save lives of other forms. The act may be regarded by him as virtuous and as one of his obligations as a physician.

As noted earlier, he may endorse the property notion of personhood. If he does, Twin *B* will not be regarded as a person, as it certainly lacks the required moral qualities. Will he save Twin *B* then? He may, on benevolent grounds, save it because he views all life forms as having a value of their own which is worth saving. He may, on humanistic grounds, refrain from saving it, as he may not see it as human and therefore as worthy of saving. Both ways of thinking seem consistent with Confucianism, depending on how generously one would allow one's benevolence to be extended. If this is the case, the Confucian person as physician seems to endorse two conflicting courses of action. Does this fact imply a defect of the Confucian conception of personhood?

In the abortion debate, the rights of the woman play a crucial role. Even if we disregard the status of personhood of the fetus in the arguments for or against abortion, the woman's right to choice cannot be neglected. Assuming he holds the genetic notion of personhood of the fetus, the Confucian person as physician may not be receptive to the rights of the woman patient and may not take the rights argument seriously. Excluding the case where the life of the mother is under threat, he may recommend saving the fetus on the ground that it is obligatory for him to save the fetus as a person. In the case where the pregnancy results from incest or rape, he may recommend saving the fetus on the ground that he puts a premium on life saving. When the life of the pregnant mother and the life of the healthy fetus are in conflict, he may recommend saving the life of the mother as the mother is already a person, while the fetus is only a potential person.

If the abortion involves a severely genetically defective fetus, he may support abortion on compassionate grounds, as he may see it not in the interest of the woman to continue the pregnancy. Again, his consideration is basically based on benevolence, not on rights. There are, however, legitimate cases where the rights of the woman deserve a hearing. As the Confucian person as physician may not be sensitive and sympathetic to rights, the rights of the woman and the rights argument in general may not be properly entertained. In the end, whether he supports or rejects abortion, the decision is based on benevolence and not on the concern for rights.

Regarding physician patient-relationships, the Confucian person will be more sympathetic to the principle of benevolence, but will experience tension with the principle of autonomy insofar as the principle invokes the rights of the agents. This does not mean that the moral person is entirely without merit. He will be more comfortable with certain version of paternalism where the rights of patients will be given a lower priority when they conflict with the welfare of the patients. He can be a fully moral person in the non-rights sense, with all those relevant virtues that a Confucian gentleman can possibly possess.

The implication of such a person *as a giver*, e.g., physician, is that his compassion and benevolence will dominate all other concerns, especially the rights of the patients. In the context of the person *as a receiver* or as a patient, he may not consider his rights to be as important as those who are rights-conscious and are willing to take a more submissive position and accept the decision made for them by the doctor. In general, a patient may voluntarily support paternalism and be willing to give up his autonomy.

In many cases in biomedicine where rights are invoked, the non-rights-based personhood may endorse a more paternalistic stance, possibly with a combination of benevolence and other virtues.

7. CONCLUSION

It has to be admitted that the rights of agents in bioethical contexts are important as they are pervasive. Similarly, the rights arguments are important arguments in many bioethical situations. A viable notion of personhood should enable us in arguments and decision making to take rights seriously. Any notion of personhood which fails to do this may be defective and inadequate.

The chief flaw of the non-rights-based conception of personhood is that it fails to take due consideration of the rationality of the rights arguments as well as the rights of the agents involved in decision-making. Without a good argument that the rights concern is immaterial or something to that effect, the failure to take rights seriously may count against that doctrine. It seems that the major weakness of Confucian personhood is its inability to take rights seriously.

Asian Research Institute
University of British Columbia
Vancouver, Canada

NOTES

[1] The meaning of a Confucian moral person referred to here is a moral person typically conceived within the Confucian moral system. The person's moral sensibilities and capacities are largely confined to the family or its institutional or social derivatives. He is highly non-social in the civic sense as he does not regard rights as central to his or others' personhood. I refer to this moral person as the personal moral person as distinct from the interpersonal moral person in the rights sense (civic type). Though the personal moral person is interpersonal within the familial sense, he is not interpersonal in the civic sense, which requires a much larger repertoire of knowledge, skills, affective capacities and duties and rights, including respecting his own rights and the rights of others.

[2] What follows is presumably the basic teaching of how man should relate to the social groupings and society that surrounds him:

> The ancients who wished to manifest their clear character to the world would first bring order to their states. Those who wished to bring order to their states would first regulate their families. Those who wished to regulate their families would first cultivate their personal lives. Those who wished to cultivate their personal lives would first rectify their minds. Those who wished to rectify their minds would first make their will sincere. Those who wished to make their wills sincere would fist extend their knowledge. The extension of knowledge consists in the investigation of things. When things are investigated, knowledge is extended; when knowledge is extended, the will become sincere; when the will is sincere, the mind is rectified; when the mind is rectified, the personal life is cultivated; when the personal life is cultivated, the family will be regulated; when the family is regulated, the state will be in order; and when the state is in order, these will be peace throughout the world. From the Son of Heaven down to the common people, all must regard cultivation of the personal life as the root of foundation (Chan, 1963, pp. 86-87).

The main focus and aim of the *Great Learning*, as epitomized by the passage quoted, was the cultivation and development of the moral self of rulers and princes as well as the common folks. Though the principal audience of this great classic was rulers and princes, its teaching could have a wider appeal. Under a broader interpretation, it presumably can be applied to the learned class or even ordinary folks.

[3] The innate goodness of man was clearly stated by Mencius:

> If you let people follow their feelings (original nature), they will be able to do good. This is what is meant by saying that human nature is good. If man does evil, it is not the fault of his natural endowment. The feeling of commiseration is found in all men; the feeling of shame and dislike is found in all men; the feeling of respect and reverence is

found in all men; and the feeling of right and wrong is found in all men... Humanity, righteousness, propriety, and wisdom are drilled into us from outside. We originally have them with us (*Mencius*, from Chan, 1963, p. 54).

[4] With the exception of the Communist era from 1949 onwards, Confucian ethics had provided the ideological foundation for the family and the state and helped to legitimize and sustain them.

REFERENCES

Berger P.L. (1983). *Secularity: West and East. Cultural Identity and Modernization in Asian Countries*, Kokugakuin University Centennial Symposium.

Chan, W.T. (1963). *Source Book in Chinese Philosophy.* Princeton: Princeton University Press.

Creel, H.G. (1970). *The Origins of Statecraft in China*, Vol. I. Chicago: University of Chicago Press.

Fairbank, J.K.(Ed.) (1957). *Chinese Thought and Institution*. Chicago: The University of Chicago Press.

Hsieh, H.W. (1968). 'Filial piety and Chinese society.' In: C.A. Moore (Ed.), *The Chinese Mind* (pp. 167-187). Honolulu: University of Hawaii Press.

Hsieh, Y.W. (1968). 'The status of the individual in Chinese ethics.' In: C.A. Moore (Ed.), *The Chinese Mind* (pp. 307-322). Honolulu: University of Hawaii Press.

Hsu, F.L.K. (1967). *Under the Ancestors' Shadow: Kinship, Personality, and Social Mobility in Village China*, revised and expanded edition. New York: Doubleday.

Hsu, F.L.K. (1971). 'Psychological homeostasis and *ren*: Conceptual tools for advancing psychological anthropology,' *American Anthropologist*, 73, 23-24.

Hsu, H-C *et al.* (1964). 'An analysis of Chinese clan rules: Confucian theories in action.' In: A.F. Wright (Ed.), *Confucianism and Chinese Civilization* (pp. 16-49). Stanford: Stanford University Press.

Ip, P.-K. (1993). 'Confucian ethics and human rights.' Paper presented at the Asian and North African Studies 34th International Congress, Hong Kong.

Ip, P.-K. (1993). 'Confucian familial collectivism and the underdevelopment of the civic person.' In: L. Lo *et al.* (Eds.), *Collected Essays in the Moral and Civic Education*. Hong Kong: Chinese University Press.

Lau, S.K. & Kuan, H.C. (1988). *The Ethos of the Hong Kong Chinese*. Hong Kong: The Chinese University of Hong Kong.

Lau, S.K. (1982). *Society and Politics in Hong Kong*. Hong Kong: The Chinese University Press.

Munro, D.J. (1969). *The Concept of Man in Early China*. Stanford: Stanford University Press.

Shelp, E.E. (1986). *Born to Die? Deciding the Fate of Critically Ill Newborns*. New York: The Free Press.

Weber, M. (1951). *Religion of China*. H. Gerth (Trans.). New York: Free Press.

Wright, A.F. (1964). *Confucianism and Chinese Civilization*. Stanford: Stanford University Press.

Wright, A.F. & Twitchett, D. (Eds.) (1962). *Confucian Personalities*. Stanford: Stanford University Press.

Wu, J.C.H. (1968). 'Chinese legal and political philosophy.' In: C.A. Moore (Ed.), *The Status of the Individual in East and West*. Honolulu: University of Hawaii Press.

CHAPTER 6

RUIPING FAN

RIGHTS OR VIRTUES?
TOWARDS A RECONSTRUCTIONIST CONFUCIAN
BIOETHICS

1. INTRODUCTION

The previous four essays in the first section of this book heuristically reflect on the foundations of bioethics in Asia. On the one hand, both Julia Tao and Edwin Hui lay out a Confucian relational view of human persons as well as its implications for biomedical practice in contrast with the Western individualistic perspective of persons. Hyakudai Sakamoto briefly argues for a distinct Asian bioethics based on the Asian cultural and moral ethos which is, according to him, fundamentally different from the background of Western bioethics. On the other hand, Po-keung Ip critically assesses the Confucian virtue-based notion of personhood and contends that it is inadequate for dealing with bioethical issues because it lacks a theory of individual rights. These two sets of essays, Tao's, Hui's, and Sakamoto's on the one hand, and Ip's on the other, represent two different approaches to the nature of Asian bioethical explorations for the foundation of an Asian bioethics. In the approach of Tao, Hui and Sakamoto, the foundation must be established based on Asian cultures, religions, and moralities, especially Confucianism, thus providing a bioethical account more adequate in the Asian context than the account offered by modern Western individualist morality. In Ip's approach, however, a rights-oriented bioethics must shape the core of an Asian bioethics, because, as he sees it, Asian moralities in general and Confucianism in particular fail to take individual rights seriously and are thereby fundamentally defective.

This essay will first draw on the arguments offered by Tao, Hui and Sakamoto to show why a rights-based bioethics is not a panacea for solving issues in Asian bioethics. Then, in response to some of Ip's concerns and arguments, this essay will lay out the strengths of a virtue-based Confucian account of personhood for bioethical explorations. Furthermore, it will compare the liberal accounts of equal rights and the Confucian account of unequal virtues as well as their respective underpinning principles. Finally, some concluding remarks are in order concerning

R.-Z. Qiu (Ed.), Bioethics: Asian Perspectives, 57-68.
© *2003 Kluwer Academic Publishers. Printed in the Netherlands.*

Reconstructionist Confucian bioethics' approach to dealing with specific bioethical issues.

2. ARE RIGHTS PERSUASIVE?

There is rights abuse in contemporary society. Influenced by modern Western individualist rights-based liberal theories, a great number of individual rights have been "created" and appealed to whenever people face social and/or ethical problems. We have found an ever-increasing amount of rights assigned to particular human, and even non-human, individuals, such as "women's rights," "children's rights," "embryo's rights," "animal rights", etc. We have also witnessed more and more special rights extended to almost every aspect of life, such as the "right to die," "right to a job," "right to a vacation," "right to a meaningful life," "right to take a nap at noon," etc. Rights have simply been taken as panaceas for solving any problem. More crucially, some have used rights as the fundamental moral standards to assess all other moral conceptions and ideas for their legitimacy. Following this tendency, Ip argues that "rights arguments are important arguments in many bioethical situations. A viable notion of personhood should enable us in arguments and decision-making to take rights seriously. Any notion of personhood which fails to do this may be defective and inadequate" (p. 55).

However, before one takes "rights" as standards to judge other moral notions, one needs at least to describe which rights one is speaking of and why they are qualified as criteria for assessing other things. Unfortunately, like many rights theorists, Ip does not intend to address these underlying issues. It seems that, to many rights upholders, rights are self-evidently clear, and so important that there is no need to offer any serious explanation and/or justification for them. They just follow Robert Nozick, and begin their arguments by assuming that individuals have rights that should not be violated by other individuals or society (1974). In the case of Ip, he immediately turns to an examination of the issue of whether Confucian moral personalism, together with its familial collectivism, suffices to build up a robust rights-based concept of personhood (pp. 53-54), without examining the issue of why an Asian concept of personhood must be "a robust rights-based concept." His conclusion is that because Confucian familial collectivism is inhibitory to rights and equality and is conceptually incompatible with the concept of a rights-based personhood (p. 54), the Confucian notion of personhood is defective and inadequate (p. 57).

I will address the Confucian notion of personhood in the next section. Here I want briefly to analyze the standing of the rights-based notion of personhood[1] that Ip takes for granted. This notion of personhood emphasizes that, among other things, a person is a bearer of rights, and must be treated with dignity and as an equal of other rights-bearers, without having his/her rights being violated by others. It takes rights as primary in the context of ethical and bioethical conflicts. However, no matter how "correct" and "important" Ip assumes this notion is, it is far from being problem-free. Although I cannot offer a systematic assessment of the rights-based notion of personhood, I will draw on some prominent arguments offered in the papers by Tao, Hui and Sakamoto to summarize a few remarkable issues relevant to the notion.

First, as Tao indicates, the language of rights cannot provide the resources for building mutual concerns and cooperative relationships between opposing parties caught in a conflicting situation of competing interests. A typical example is abortion: the fetus' rights are easily opposed to the pregnant women's rights to pursue her own life plan. Moreover, the father, the grandparents, and the state may also contest the woman's right to terminate or to continue her pregnancy. What becomes clear is that the language of rights and the legal system based on it tend often to exaggerate rather than reduce the division between the different parties involved (Tao, p. 15).

Second, the primacy of rights tends to obscure the appropriate relation between individuals and society. It tends to overemphasize concerns with individual liberty and self interest, seeing the self as essentially separate from others. In this way, it may lead to extreme positions and actions. For instance, in health care allocation, stressing health care rights can lead to an unlimited demand on government provisions to satisfy maximum individual preferences, resulting in escalating costs and the unsustainability of the entire system. It can also lead to the gradual erosion of the collective provision of public health care because of increasing demand for freedom of choice and individual responsibility for health care through the market and private insurance. Either way can bring about the gradual disappearance of public health as a common good in society (Tao, p.16).

Third, while the primacy of rights leads to emphasizing individual autonomy, free choice, and self-determination, a major difficulty of such an emphasis lies in its underlying notion that individuals can be abstracted from relationships, social contexts, and even from qualities of human agency that are vital to human life, namely the capacity and need for connectedness, relationships and mutual care. It tends to reinforce separation and isolation, marginalizing family involvement and shared family-determination. For instance, most discussions of euthanasia and advance directives in past decades have been framed in terms of protecting the autonomous choice of patients to decide when to die. However, despite this autonomy, we may feel we really have no choice, largely because we are unable to find meaning in death or bring our lives to a meaningful close. We suffer from an inability to give meaning to death because death defined as an individual right and autonomous choice is isolated, disconnected, and dislodged from the web of personal relationships and community bonding that give meaning to life and death (Tao, p. 19).

Finally, as Hui demonstrates, the rights-based notion of personhood is developed from the long-standing Western view of "substance." One of the most crucial characteristics of the concept of substance is its aptitude to exist in itself as a concrete individual thing and not as a part of any other being. The philosophical circles of the modern West have largely retained this "substantialist" tradition (p. 31). This substantialism, as Sakamoto explains, was fortified in modern times by the idea of "person" and "human dignity." "Person" is identified and dignified as a rational being, assigned with the so-called "human rights." In this regard, American bioethics can be taken as an attempt to establish a legal system to deal with bioethical issues from the point of view of individual human rights. A human person is primarily a rational individual substance. This assumption of rational substance contributed to contemporary bioethics the narrowly psychological and individualistic understanding of personhood and the rights-based ethic. These are sharply different from the Chinese ethic of personhood that has been developed from the Confucian moral tradition.

This is to say, in the understanding of Tao, Hui and Sakamoto, the rights-based notion of personhood has been developed from particular Western metaphysical and moral traditions and is characteristic of specific Western moral contexts. It is by no means universally acceptable or without need of further discussion. More importantly, as Tao, Hui and Sakamoto see it, this notion, with its emphasis on the primacy of individuals' rights, often causes more serious confrontations between agents than would attempts persuasively and harmoniously to solve problems. Accordingly, it is inappropriate to take this notion of personhood for granted and as a standard for measuring all other notions of personhood as if it were uniquely perfect and universally justified.

3. THE CONFUCIAN VIRTUE-BASED PERSONHOOD

The Confucian notion of personhood is a virtue-based notion.[2] Unlike the contemporary concept of rights, which assumes various individual entitlements based on a faith in universal reason to "intuit" such entitlements, independently of any particular understanding of human nature, a concept of virtue is inherently relevant to a specific view of human nature. Generally, virtue is understood as good character. It is not only a state of affairs, but is also proper human activity. A coherent set of virtues must be a complete series of human traits or qualities that enable humans to do the right thing at the right time in the right way in pursuing the good life suitable for human nature. Hence, a particular theory of virtue cannot be established independent of a specific conception of human nature. For instance, Aristotle's theory of virtue is based on his understanding that man is by nature a political animal (neither a beast nor a god), possessing the potential to form the (just) polis. In contrast, the Confucian theory of virtue is based on its view that man is by nature a familial animal, possessing the potential to form the (appropriate) family.[3]

The Confucian understanding of the nature of humans as familial animals involves a deep idea of love. It traces back to the original work of the Confucian master, Confucius (551-479 BCE), especially in his reconstruction of the Chinese rituals (*li*) in terms of the fundamental Confucian virtue, *ren*. Confucius lived in a time of immense social conflict and turmoil in Chinese history. By his time, China already had thousands of years of civilization with the glory of its ritual system. But the glory was declining in his time. The rich and powerful feudal lords, from a time of unprecedented economic growth, wanted to grasp more political power. The states began to wage wars against each other, subjects murdered their princes, and children killed their parents. The ritual system was disintegrating, and the morality of the society was degenerating. Confucius, among a dozen of the most brilliant minds that China had ever produced, reflected on human nature and looked for a comprehensive strategy to reform the society. He was not unique in recognizing that humans must form appropriate behavior patterns and follow proper rules in order to live together peaceably. But it was truly significant for him to vindicate the idea that appropriate behavior patterns and proper rules are exactly illustrated by the traditional Chinese rituals, and that such rituals should be restored and maintained in order to put society back on the right track.

Confucius did not mean to keep intact all behavior patterns or rules implicit in the established rituals, though. Rather, he was to reconstruct the Chinese ritual system in terms of a fundamental human virtue that he teased out through his reflection on human nature. Humans, for Confucius, are not atomistic, discrete, self-serving individuals coming to construct a society through contract. They are first and foremost identified by the familial roles that they take on: husband, wife, father, son, mother, daughter. Every human must be born and grow in the family. This familist way of human existence is not chosen, but rather is given. The parent-child relation not only becomes the most important human relation, but it also vividly illustrates a most significant and noble aspect of human nature: humans' ability to sympathize with each other (that is, one cannot bear the suffering of the other). This natural sympathy constitutes the human disposition of love. For Confucius, the love between parent and child sets down the root of a fundamental human virtue, *ren* (*Analects* 1: 2), and this virtue must be cultivated and promoted in order to build a good society. Thus, *ren*, among a number of the traditional Chinese virtues, was teased out by Confucius as the fundamental human virtue to reconstruct the ritual system.

Indeed, Confucius clearly articulated *ren* as loving humans in the *Analects* (12: 12). In this general meaning, it seems that *ren*, upheld by Confucius as the fundamental human virtue, has no essential difference from the word *ren* that had already been used in the pre-Confucian literatures, such as the *Classic of Songs* (*Shi Jing*) and the *Classic of Doctrines* (*Shu Jing*). However, what is special about this virtue is the foundation that Confucius set down for it so as to reinterpret it as a profound and complete human virtue. Evidently, love needs an impetus. In other words, love must have a foundation to empower itself. The foundation of Christian love is a creative tie between God and man; the foundation of Buddhist love is a causal connection among all beings; and the foundation of Ancient Greek love is a romantic union between man and woman (or man and man). Unlike any of these, the foundation of Confucian love is a blood-tie between parent and child. Confucius must have been extremely impressed by the power of this parent-child love when he used it to account for the root of the fundamental human virtue, *ren* (*Analects* 1: 2). Parent-child love is not romantic love. It engages a deep affection towards and selfless care of each other, having nothing to do with mutual sexual attraction or impulses. It is not even reciprocal – the parental love of one's children or the children's love of their parents is not the granting of privileges in return for similar privileges from the other side in the future. It is nonsensical to say that they are making mutual concessions with each other for self-benefit. The meaning of their love transcends any sense of contract. Finally, for Confucians, only through the establishing, nurturing, and developing of the parent-child love in the family and gradually extending it to other people outside of the family can a good society be possible. Hence, love must begin from the parent-child tie inside of the family. If it cannot begin from the family, it will begin from nowhere.

Moreover, the virtue of *ren* as love should not be taken as only a feeling of love. Reading the *Analects* carefully, we can see that *ren* is foundationally a potential, a power, and a character that humans possess, from which the feeling of love naturally arises. In this regard, Mencius (327-289 BCE), the Confucian master second only to Confucius, is quite right in insisting that everyone is born to have a seed of *ren* in their heart/mind – the seed has both the substance and function (feeling) of love. Without the substance, there would be no function. But Mencius might have misleadingly gone beyond Confucius in claiming that human nature is thereby good. This assertion makes some people mistakenly conclude that they can become virtuous without the need to follow any guidance or make an effort. This is certainly not what Mencius really meant. A seed is not a fruit. The fundamental virtue as a good and noble aspect in human nature is only a potential, which needs cultivation, nurture and development to fully realize itself. This is why Confucius emphasized the importance of following the rituals in order to pursue virtue (*Analects* 7: 1). It turns out that, in Confucianism, *ren* (as the fundamental human virtue) and *li* (as the proper human behavior patterns and rules) hold the dialectical relation of mutual contribution and restriction.

In the first place, the purpose of establishing and maintaining *li* for human society is nothing but contributing to nurturing and promoting the fundamental human virtue, namely *ren*. Without this purpose, *li* is simply pointless (*Analects* 3: 3). *Li* should not be understood as something like artificially man-made laws imposed on humans from outside. It is rather a net of indispensable roads directing one to the good life suitable for human nature. Hence, Confucius recognized how important it is to follow *li* for pursuing virtue. In fact, in his view, the only way for a human to become a human of *ren* (namely, to realize his/her nature fully) is by subduing his/her passions and complying with the established *li* (*Analects* 7: 1). He clearly taught that the method of realizing *ren* is "not to look at what is contrary to *li*; not to listen to what is contrary to *li*; not to speak what is contrary to *li*; and not to make a movement which is contrary to *li*" (*Analects* 7: 1).

On the other hand, however, it is not Confucius' position that every particular *li* is always appropriate and should never be revised. For him, the *li* of a new time is unavoidably a reconstruction of the *li* of the old time. As he stated, 'the Yin dynasty built on the *li* of the Xia. What was added and what was omitted can be known. The Zhou dynasty built on the *li* of the Yin. What was added and what was omitted can be known" (*Analects* 2: 23). Importantly, for him, reconstruction must be made according to the fundamental human virtue in consideration of new circumstances. When a disciple asked about the basis of *li*, Confucius answered: "a noble question indeed! With ceremonies, it is better to err on the side of frugality than on the side of extravagance; in mourning, it is better to err on the side of grief than on the side of formality" (*Analects* 3: 4). This is to say, the essence of following *li* lies more in your internal sincerity that embodies love than in your apparent behaviors that conform to the rules. Confucius gave an example to show how he would revise a rule of *li*: "a ceremonial cap of linen is what is prescribed by *li*. Today black silk is used instead. This is more frugal and I followed the majority" (*Analects* 9: 3). He also gave an example to show why he would not change a rule of *li*: "the *li* prescribe the bowing below the hall, but now the practice is to bow only after ascending it. That is arrogant. I continue to bow below the hall though going against the majority" (*Analects* 9: 3). That is to say, the majority view may not be the true standard of *li*. The true standard is in conformity with the conception of *ren* – the rules and rites that contribute to the substance of love should be established and maintained regardless of contemporary common practices.

In short, for Confucians, the fundamental human virtue is *ren*. All humans possess the seed of this virtue and it must be nurtured, developed, and perfected in order for each to become a true human person and realize human excellence. Confucius reconstructed the Chinese rituals in terms of this fundamental virtue, *ren*. (1) The rituals (as a system of behavior patterns and rules) are morally meaningful because they are indispensable to the fulfillment of the virtue.

(2) Everyone must follow the rituals in order to nurture and develop their virtue.

(3) Rituals can be revised according to the central concern of the virtue.

These Confucian understandings provide, among other things, a virtue-based conception of the human person. Whatever a human is, he/she is a virtue-holder and pursuer. A human is a virtue-holder because he/she has been invested with the potential of loving his/her parents and children and capable of extending this love to other people. A human is also a virtue-pursuer because he/she must nurture and develop this potential in accordance with the requirements of rituals in order to become a virtuous human. In comparison with the rights-based notion of personhood, this Confucian virtue-based notion of personhood is (1) grounded in a blood-tie-based natural love between parent and child (rather than on emphasis on universal reason) so that people are naturally driven in practicing this notion of personhood, (2) duty-oriented (rather than claim/entitlement-oriented) so that it carries more mediating power in dealing with human problems and conflicts, and (3) family-oriented (rather than individual-oriented) so that it promises natural human unions and harmonious relationships.

4. REFLECTIONS ON EQUAL RIGHTS VS. UNEQUAL VIRTUES

Ip is right in stating that filial piety (namely, the manifestation of children's love of their parents) becomes crucially important in pursuing the fundamental Confucian virtue of *ren*. Indeed, compared to the parental love of their children, the children's love of their parents needs more cultivation and promotion for its exercise. This is why Confucians often emphasize more of the children's care of their parents than the parents' care of their children. Ip contends that the emphasis on filial piety leads to a hierarchical human family in which equal rights have no place. The hierarchical and paternalistic familial and social structure successfully preempts equals among persons. The emperor, the father and the husband occupied norm-generating and norm-dictating statuses and positions. All other social relationships and institutions would spin around these unequal and domination-*cum*-subordination social arrangements (p. 57). Hence, the Confucian society as well as the Confucian family does not have the idea of treating everyone as an equal. Confucian persons do not have equal rights. This, for Ip, indicates that the Confucian view is defective.

It is true that Confucian familism does not hold the liberal democratic principle that people should be treated as equals.[4] Instead, in my view, the Confucian principle is that people should be treated as relatives. As I indicated in the previous section, from the Confucian understanding of humanity, separate individuals cannot become true humans. It is the relationship of love among humans that produces and identifies the good character of human individuals. The root of this love is foundationally grounded in the parent-child tie and intimate affection. It is this love that brings out the basic human community, namely the family, and makes possible the creation of a civilized human society, insofar as this love can be nurtured, promoted, and extended gradually to others outside of the family. Hence, the feature of this love is inconsistent with any basic egalitarian moral sentiment, and the basic structure of the Confucian model of the family (on which all other human relations and social institutions are modeled) is at odds with the liberal principle of treating people as equals.

Although the principle of treating people as equals does not necessarily support the equal distribution of income or wealth, it essentially requires equal respect of everyone and equal consideration of everyone's interests. It sets down the basic tone of the "equal rights" that liberalism emphasizes to the utmost. However, Confucianism does not accept equal respect of everyone or equal consideration of everyone's interests. To be sure, Confucians understand that each human, *qua* human, possesses the seed of the virtue, *ren*, and is able to cultivate and practice love, first inside of the family and then gradually extended to outside of the family. In this sense every human is able to love and respect others and in turn deserves others' love and respect. However, Confucians also recognize that human individuals differ from one another in the degree in which they learn and exercise the virtue in respecting and loving others. Because some are more sincere in attitude, make more efforts in act, and accomplish more achievements in outcome than others in the practice of the virtue, Confucians hold that they deserve more love and respect (and their interests should be considered more) than others in society. This "unequal respect" constitutes one of the fundamental Confucian values: *zun-xian* (honoring the men of virtue). The Confucian classic *Doctrine of the Mean* regards *zun-xian* as one of the most important moral standards in regulating society (20: 13 – 14). From the view of the *Doctrine of the Mean*, by honoring men of virtue, society can be preserved from errors of judgment.

Some might want to argue that even if the Confucian principle of treating people as relatives sounds unegalitarian, it still contains an egalitarian thesis that can be dug out. This is because, they might contend, there must be a "threshold" in treating individuals as relatives below which they are no longer being treated as relatives. This is true theoretically. However, there has never been such a threshold that has been established and emphasized by Confucians in order to use it as a general and universal standard for treating all individuals equally or similarly. Instead, with the basic spirit of loving and caring for everyone, Confucianism always gives attention to the different roles, characters and circumstances of different individuals. Importantly, based on the Confucian understanding of human nature, humans are not mutually disinterested and independent individuals as John Rawls assumes they are in the so-called "original position" (1971). Rather, they are always familially and socially related to each other, and their relations are bound to be asymmetrical: some are born as intimate family members, some are made close friends, and some are only remote strangers. Although the Confucian virtue of *ren* requires extending love from one's family members to all others outside of the family, it does not hold that one should love everyone equally or similarly. To the contrary, Confucianism always requires that there ought to be a clear and definite order, distinction, and differentiation in the application of love (Chan, 1955, pp. 8-9). It is not that one should not love all; it is that one should love all with differentiation and relativity of importance. This peculiar Confucian discrimination is well reflected in the Confucian slogan of "love with distinction" or "care by gradation" in the process of Confucian self-cultivation. It constitutes another crucial Confucian value: *qin-qin* (affection towards relatives) (see, e.g., *the Doctrine of the Mean* 20: 13-14). It is wrong for one equally to consider one's close family members' interests and strangers' interests. It is only appropriate to give more weight to the interests of one's family members. Importantly, this unegalitarian moral ethos has to be embodied in public and social policy. For instance, welfare is basically a family responsibility. It is mistaken for the state to offer comprehensive welfare programs in order to equalize everyone's welfare in society. Such egalitarian attempts are in tension with the basic Confucian familist moral sentiment because they deprive families of their own resources from seeking welfare for their family members as they see best.

As the family is the basic Confucian pattern of normal human existence, equal rights in the family is a non-starter for Confucians. For the healthy growth of children, what is most important is not to respect their liberty, free choice, or self-determination, because these capacities have not yet been developed in children. What is most important, rather, is to instill children with appropriate rules, knowledge, and patterns of behavior in a way that children find interesting to learn so as to cultivate them to become virtuous, normal, human adults. For the care of the elderly in the family, what is most important is not to respect their rights as "equal" to the rights of others. Rather, it is to take care of their special material, emotional, and spiritual needs so as to help them lead a happy elderly life (Wang, 1999). In short, from the Confucian view, "unequal virtues," rather than "equal rights," is the more pertinent feature of the family. Rights are not sufficient for proper human concern and care.

On the other hand, it is misleading to believe that Confucianism supports dictatorship in human relations. It is true that Confucians hold that the father, the husband, and the king should play more active roles in the relations of father-son, husband-wife, and king-subjects so as to maintain the normative and harmonious relations according to Confucian concerns and commitments. But it is a misunderstanding to think that they are in the position of articulating the norms and standards of morality bearing on human relations. The norms and standards are already set down by the Confucian understanding of the fundamental human virtue, *ren*, and formulated by the rules of the ritual system, *li*. To understand Chinese history, it is crucially important to distinguish what is taught in the Confucian classics and what was realized in the actual politics of the imperial courts. The imperial dictators always managed to distort the Confucian teachings to serve their own interests under despotism. *Pace* Ip, the true spirit of the Confucian relations, as Tao points out, is not domination-cum-subordination, but is reciprocity. Reciprocity is not equality. It is virtue-based interconnectedness, interdependency, and interactions (p. 24). As Hui indicates, the moral development of the individual in the authentic Confucian tradition is not the establishment of the independence or individuation of the individual, but the maintenance and promotion of the harmony of the community (p. 36). The Confucian social ideal is a harmonious human community based on the virtue-based notion of personhood, in which everyone is taken care of according to his or her particular circumstances.

5. TOWARDS A RECONSTRUCTIONIST CONFUCIAN BIOETHICS

From Ip's point of view, a virtue-based Confucian bioethics may not be very helpful in dealing with complex bioethical issues, such as abortion, euthanasia, organ transplantation, human experimentation, and artificial human procreation. In addition to pointing out that Confucian physicians may not be sensitive and sympathetic to rights, such as women's rights (p. 58), he also suggests that Confucian virtue theory may not be able to give unambivalent guidance to all specific bioethical matters, such as what the physician should do with a severely defective newborn (p. 58).

It is not always easy to apply a general ethical theory to a particular ethical case, because different understandings and descriptions of the case may occur even when the theory itself is clearly formulated and interpreted. The substance of the theory may also be explained in different ways by different theorists. In other words, scholars may often hold different interpretations of an ethical theory. I admit that this is the case for the Confucian virtue theory – not all Confucians agree about what the substance of the theory exactly is. Therefore, not all Confucians will come to the same solution to a particular bioethical case or issue. However, this is not unique to the Confucian theory. It is even more so for human rights theory. Rights theorists hold all sorts of disagreements regarding who should have rights, what rights one should have, and what the proper scope of a right is when one has the right. Due to such disagreements, rights theorists can never give a unanimous answer to a bioethical question based on their theory.

Still, the basic differences between the liberal rights theory and the Confucian virtues theory are as fundamental as they appear. We need to engage in a careful comparative philosophy to explore which is more persuasive for bioethical practice in Asia, without taking one of them for granted before ethical deliberation. In fact, at the beginning of this new century, many Asian people have acquired a sense of moral deportment, bioethics, and health care policy that are radically different from individualist liberal ethics and politics predominant in the West. No doubt, this difference can be traced to numerous sources. But it is both challenging and significant to restructure properly-understood Confucianism in order to show how this difference can serve as a cultural and moral resource. I attempt to approach this challenge through a proposal for the reinvigoration of Confucian thought under the rubric of Reconstructionist Confucianism (Fan, 2002).

We shall have a great deal to do in Reconstructionist Confucianism. First, we should recognize that all past attempts to read into Confucianism liberal democratic concerns with liberty, equality, human rights and liberal democracy (either by claiming that Confucianism already held such values or by arguing that Confucianism can, through certain necessary transformation, easily produce such concerns), is a form of naïve presentism. The truth of the matter is that cardinal Confucian concepts (such as *ren* and *li*) presuppose understandings of morality and justice that are not reducible without loss of essential meaning to liberal individualist concerns with equality, rights and justice. In this regard, Ip should be credited for sharply articulating the basic tension between the Confucian virtue-based morality and the liberal rights-based morality. But he is too quick to claim that the former is internally defective and thereby inferior to the latter.

Second, Reconstructionist Confucianism holds that it provides a more ample account of human flourishing and morality than that offered by other accounts, individualist liberal accounts included. This account is not individual-oriented, equality-oriented, or rights-oriented; rather, it is family-oriented, *junzi*-oriented, and virtue-oriented. The individual is encouraged to live out a Confucian way of family life through cultivating virtue and becoming a *junzi*, an exemplary person of good character and moral integrity. In this regard, Confucian scholars need to engage in a detailed communication with liberal scholars so as to disclose their profound disagreements and explore a possible contemporary Confucian philosophy for practice. For this mission, Confucian scholars must turn to particular ethical and social issues (such as controversial bioethical issues) to offer their solutions based on basic Confucian values and commitments. They cannot achieve this goal by simply staying in the ivory tower to explore "theoretical" issues without touching reality.

Finally, Reconstructionist Confucianism is meant to reconstruct authentic Confucian community in Asia according to the fundamental Confucian virtue in consideration of contemporary social circumstances. It calls for restructuring social institutions through reformulating public policy in accordance with fundamental Confucian moral and political commitments rather than modern Western liberal social-democratic concerns. The success of this enterprise will not only determine the fate of Confucian civilization, but also determine the extent to which the Confucian community will eventually contribute to the world civilization.

City University of Hong Kong
Kowloon, Hong Kong

NOTES

1. There are different notions of personhood in terms of rights consideration. In another paper (2000, pp. 20-23) I identify five different conceptions of personhood that I term "appeal to rights" conceptions. They have offered five different specific standards for measuring personhood: the standards of species, potentiality, sentience, brain or heart functioning, and awareness of self as a continuing entity. For the sake of simplicity, I do not in this paper mean to engage in the details of these different rights-based or "appeal to rights" conceptions of personhood. I take all these conceptions to be based on a particular moral point of view: a person, whatever he or she is, is a rights-holder. In this sense, I will simply talk about "the rights-based notion of personhood" in the text as Ip does in his.
2. In my (2000), I identify the Confucian conception of personhood as an "appeal to rites" conception in contrast with the Judeo-Christian "appeal to creation" conception (pp. 16-20).
3. In the Confucian view, the family is the foundational unit of society, and society is nothing but an enlarged family. For the different understandings of the family and human relations between Aristotelianism and Confucianism, see my (2002).
4. For a clear articulation and explanation of this principle, see, e.g., Dworkin (1978). All leading liberal arguments can be taken as based on this fundamental principle. See, e.g., Kymlicka (2002, pp. 3-4).

REFERENCES

Chan, Wing-tsit (1955). 'The evolution of Confucian concept Jen,' *Philosophy East & West*, 4(January), 295-315.
Confucius (1971). *Confucian Analects, The Great Learning & The Doctrine of the Mean*. James Legge (Trans.). New York: Dover Publications, Inc.
Dworkin, Ronald (1978). *Taking Rights Seriously*. Cambridge: Harvard University Press.
Fan, Ruiping (2002). 'Reconsidering surrogate decision-making: Aristotelianism and Confucianism on ideal human relations,' *Philosophy East & West*, 52(3), 346-372.
Hui, Edwin (2003). 'Personhood and bioethics: A Chinese perspective,' in this volume, pp. 29-43.
Ip, Po-Keung (2003). 'Confucian personhood and bioethics: A critical appraisal,' in this volume, pp. 49-57.
Kymlicka, Will (2002). *Contemporary Political Philosophy: An Introduction*, 2nd ed. Oxford: Oxford University Press.
Mencius (1970). *The Works of Mencius*. James Legge (Trans.). New York: Dover Publications, Inc.
Nozick, Robert (1974). *Anarchy, State and Utopia*. New York: Basic Books.
Rawls, John (1971). *A Theory of Justice*. Cambridge: Harvard University Press.
Sakamoto, Hyakudai (2003). 'The foundations of a possible Asian bioethics,' in this volume, pp. 45-48.

Tao, Julia (2003). 'Confucian and Western notions of human need and agency: Health care and biomedical ethics in the twenty-first century,' in this volume, pp. 13-28.

Wang, Qingjie (1999). 'The Confucian filial obligation and care for aged parents.' In: R. Fan (Ed.), *Confucian Bioethics* (pp. 235-256). Dordrecht: Kluwer Academic Publishers.

PART II:

BIOETHIS IN ASIAN CULTURE: GLOBAL OR LOCAL?

CHAPTER 7

ANGELES TAN ALORA

PHILIPPINE CULTURE AND BIOETHICS

In the Philippines people often dare not behave as they might like if such behavior is not acceptable to other people. We are not ready to face the opposition it would raise. For instance, if I am asked to deliver a presentation at a time I very much prefer to take siesta, for the sake of *pakikisama* (smooth interpersonal relations) I would not request a change of schedule. Such is our culture.

In what follows I shall try to disclose, explain and discuss how Filipino culture affects behavior and focus on how it relates to bioethics. I shall discuss definitions and their importance, describe common Filipino behaviors and their effect in the application of bioethical principles, the issues that arise in such contexts, and finally, recommendations.[1]

1. CULTURE

Culture is the way of life of a people (Gorospe, 1988, p. 87). It is the body of customary beliefs, social forms and material traits constituting a distinct complex of tradition of a religious, racial or social group. It is the complex of ways which a society develops in response to life.

1.1 Importance

Culture touches on every aspect of the ordinary life of the people, including health care. Where one gets sick determines the treatment one will receive (Miranda, 1994, p. 21). My classmate was waiting in an airport in the United States of America for a flight home when she developed mild chest pains. She was rushed by ambulance to the nearest hospital, immediately had an angiogram, and had a coronary bypass within 24 hours and before any family member arrived. A colleague had a similar pain in an airport in Manila. She was put in a taxi by another passenger, then accompanied by her daughter to the clinic of a cardiologist. It took a week before cardiac studies were completed and another 3 weeks before surgery was performed (the doctor she chose was going abroad and preferred to do the surgery only on his return). By this time, the whole family had been informed and had agreed to the

R.-Z. Qiu (Ed.), Bioethics: Asian Perspectives, 71-81.
© 2003 Kluwer Academic Publishers. Printed in the Netherlands.

procedure. They had rescheduled their own activities so that there would always be someone with her as well as someone to take over her responsibilities.

Scientific medicine determines certain facts, whereas culture determines beliefs about such facts and ethics determines the conduct towards such facts. There is a need therefore to assess Filipino bioethics with issues, analytical tools and practical solutions derived from our own culture (Miranda, 1994, p. 21). A Filipino bioethics cannot, nor should it be, identical with Western bioethics, for bioethics is differently construed and interpreted depending on one's notions of health, illness and treatment, of life and death, and the perspective of doctor, patient, third person and observer, depending on what is available in one's setting.

What is bioethically correct in the Western sense may not be culturally acceptable. What is culturally the norm may not be bioethically correct. We must consider this interplay to determine what is both culturally and bioethically acceptable.

1.2 Assumptions

Filipino culture is difficult to define; what we commonly see here may really be common to all Asian countries or even to the rest of the world. We have been influenced by our many colonizers, visitors and the media so that what is uniquely Filipino is almost non-existent. What is attempted here is the description of common behaviors: "psst" heard in a foreign country is usually a Filipino calling another person, a diner using a spoon to cut his meat is usually a Filipino diner. Generalizations will be made, but by no means do we presume that these are absolute or that there are no exceptions.

Additionally, aspects of our culture vary in degrees in different regions, at different eras, and among different ages. The men from the Bicol areas supposedly make good husbands and the women from Pampanga servile wives. Rural people are more respectful than those from urban areas. The culture of five centuries ago may be very different from that of today; tradition is overcome by adaptability; the elderly are more traditional than the young.

Finally, it must be noted that culture has both good and bad effects on behavior.

1.3 Characteristics

The characteristics we shall discuss as common in the Filipino culture are social corpus or family orientation, search for harmony, person orientation, and multidimensional health care.

2. FAMILY ORIENTATION/SOCIAL CORPUS

The Filipino is clannish. The family is considered "the highest value in Filipino culture" (Andres, 1981, p. 22) and the core of all social, cultural and economic activity. The family provides financial and psychological support, emotional security, and a feeling of belonging – an environment where a Filipino can be himself.

For the individual Filipino, all basic needs, food, clothing and shelter, come from the family. Family responsibility includes mutual assistance, group advice and joint decision making. The stronger members help the weaker members; the richer, the poorer. The aged, the infirm, and the young are cared for by the family. Unmarried members continue to be supported financially and live in the family home. Whether at school or in work, family members take precedence over others (Andres, 1981, p. 110).

The family is a Filipino's defense against a potentially hostile world, an insurance against hunger and old age (Andres, 1981, p. 110). There are always relatives to whom the Filipino can turn. Receiving aid from relatives when in material need or being given advice and support in times of conflict is normal. Intruding into a relative's affairs involves deep concern for the person, a legitimate expectation (Andres, 1981, p. 112).

Important decisions such as career and marriage choices involve the entire family. Every competent adult participates in the weighing of alternatives, with the head of the family as the final decision maker. Paternalism is the rule (Andres, 1981, p. 112). All share the benefits and burdens resulting from the decision.

A Filipino's self-esteem rises and falls with that of the family members (Andres, 1981, p. 110). The decision as to whom to marry is not made isolation; when one marries someone, he marries her family (Gorospe, 1988, p. 133). One takes special pride in being identified with a family that has a good reputation – in introductions we say "*anak ni*" ("child of"). On the other hand one feels shame in being identified with a family of ill repute. Success or failure is not individual, but family-oriented. The most painful ignominy for a Filipino is to be disowned by his own family.

Early in childhood the Filipino is taught to cultivate goodwill, to be loyal to his family, and to know who is his relatives are (Andres, 1981 p. 110; Miranda-Feliciano, 1990, p. 77). He is told "*lolo mo iyan*" ("he is your grandfather").

The support and belonging that results from the close family involvement is usually helpful. It helps the permanence of marriage and the stability of its members. (Gorospe, 1988, p. 24). Its interdependence, however, also leads to a lack of independence and self-reliance (Gorospe, 1988, p. 24). When an individual Filipino has to make his own decisions, he often is not capable of trusting his own judgment and feelings. Even if able to do so, if this decision is contrary to that of his family, tension arises. He is expected to give up his own rights (To Thin Ahn, 1994, p. 11; Andres, 1981, p. 111). The same happens when he tries to assert his privacy.

When sickness occurs, the patient is a family member in need. The whole family is involved in her illness (Buenavista-Tungpalana, 1997, p. 86) Her family members rally to provide for her needs, relieve her of responsibility and exempt her from the stress of decision making. Business and home matters are taken care of by other family members. If she is the head of the family, she temporarily loses her position of authority and another family member takes over. She is admonished not to worry about anything but to think only of resting and getting well. The patient is accompanied by a family member when she goes for consultation, she calls home before following advice for hospitalization and hospitalization bills are often shared by other family members. The second clinic chair and additional hospital bed for a patient's *alalay* (carer; usually a relative) or *bantay* (watcher) is evidence of this relationship. This *alalay* acts as her aide, adviser and interpreter, helping her move around, giving her history and participating in decision making. In serious illness, this relative is often informed of the patient's condition before the patient herself.

Confidentiality does not exclude relatives from receiving a great deal of private information concerning the patient. This is justified by the fact that health care decisions are family decisions, and by the need for mutual support and help. Even the physician is often unwilling to make the patient assume a more active role in her illness. Autonomy takes on a different meaning in such a context.

The sick person accepts this role of limited freedom, dependency and passive tolerance. She feels assured that others will take care of and support her, especially if she is terminally ill. She does not mind having other people know of her illness, for it will mean more people will visit her and provide support. She does not resent visitors with inquiring questions; they are concerned about her.

The social corpus of the family progressively extends in the Filipino setting from immediate relatives to distant relatives, godparents, neighbors, townmates and nation mates (To thin Ahn, 1994, p. 86). "*Kababayan*" (nation mate) is a strong term; "*tita*" (aunt) and "*ate*" (older sister) are terms of respect.

3. HARMONY

Related to close family ties (because it fosters the social corpus) but extending beyond it is harmony. Harmony is unity, peace and cooperation in one's relations with one's self, other persons and one's environment.

Harmony is maintained through *pakikisama* (smooth interpersonal relationships), non-confrontation, *hiya* (shyness, delicacy of feeling shame or guilt), submissiveness to leaders, *utang na loob* (internal debt of a gratitude), and *bahala na* (come what may).

3.1 Pakikisama

> To be secure, the Filipino needs a sense of belonging to a group of his own kind. The price of security is loyalty to one's "in" group. It demands *pakikisama*, which is an attitude of give and take, a sensitivity to the feelings of others (Gorospe, 1988, p. 32). The Filipino learns to subordinate his own personal desires, convictions and standards to those of his group, be it family, clan, social group or *barkada* (gang) (Miranda-Feliciano, 1989, p. 22; Andres, 1981, p. 17).

Maintenance of harmony by "giving in", by practicing tolerance and consensus within family or among peers takes precedence over justice and honesty. A Filipino may agree to do something he would otherwise choose not to do in order to please the group. A family member might avoid reporting a sibling's irrational behavior, a student might do his classmate's assignment or allow his classmate to copy in a test, because of *pakikisama*.

Because of *pakikisama*, a health care provider might participate in a procedure he would otherwise not have performed, and a health care professional might not call the attention of an authority to misdeed or error of a colleague.

3.2 Non-confrontation

In order to prevent confrontation, a Filipino tries to be non-offensive even when he disagrees or refuses. When invited to a party, he answers "I will try" rather than "I can't or won't come". Harsh, insulting and harmful words and gestures are avoided. Bad news is either not told, or is told in less grave terms or disclosed in euphemisms (Andres, 1981 p. 18).

A patient is seldom told she has a serious illness. Breast cancer is *bukol* (lump), lung cancer is "tuberculosis", and liver cancer, "hepatitis". A patient is rarely informed that she is dying. Decisions regarding terminal care are therefore seldom guided by informed consent.

3.3 Hiya

Depending on its context, *hiya* means shyness, a delicacy of feelings, embarrassment, shame or guilt. It is a painful emotion arising when relating to an authority figure or society with strict norms. The Filipino is *nahihiya* because others may expect much from him (Miranda-Feliciano, 1989, p. 40). He inhibits self-assertion in a situation perceived as dangerous to his ego (Gorospe, 1988, p. 31). He may not agree to perform, for example singing or dancing in front of strangers, because he is *nahihiya* (shy).

Hiya as a delicacy of feeling, propriety and good manners means that a visitor will not come during siesta hour (Miranda-Feliciano, 1989, p. 40), and a subordinate will not call his superior on the telephone.

Walang hiya means the absence of this inhibition. It offends the finer feelings of others (Gorospe, 1988, p. 34). So a Filipino may repress disruptive feelings – discontentment, anger, contempt, or hate – and instead withdraw or behave hypocritically (To Thin Ahn, 1994, p. 4).

When one is rejected or insulted one is *napapahiya* (embarrassed). To avoid this, a student might hesitate to ask the teacher to explain a difficult concept, and an employee might hesitate to ask an employer for a favor. Similarly, when one performs a shameful act (for example, giving a wrong answer, failing to keep one's word (Miranda-Feliciano, 1989, p. 40), being unable to repay a debt, or in some way violating a norm (Andres, 1981, p. 18)), there is loss of face or defamation or one's self-esteem. One is *napapahiya* (shamed) (Miranda-Feliciano, 1989. p. 41). This may result in a person going to great lengths to avoid wrongdoing, to avoid getting caught (Gorospe, 1988, p. 66), or to cover up a wrong that has been done. He lives in the excessive fear of losing face. If the wrongdoing cannot be avoided then *hiya* becomes guilt and may lead to depression and suicide.

Because of *hiya*, a patient will not question a doctor regarding medical terms he does not understand, or actions he does not agree with. She would rather suffer in silence than risk *hiya*. A nurse, intern, or resident will not call attention to a confusing consultant's order. Erroneous acts are "covered up". All these undermine a patient's best interest.

3.4 Submissiveness to "Leaders"

Submissiveness to leaders is related to both *pakikisama* and *hiya*.

The Filipino is concerned about what "leaders" – be they heads of the family, persons with status, authority figures, or others higher in the hierarchy system – think about him. He is prone to shape his behavior accordingly, and tends to reveal to these persons only those aspects of himself that are gauged acceptable to them. In the presence of these leaders, he is respectful, a good listener, unwilling to challenge and reluctant to speak his mind. He looks up to them for support and willingly lets them make the decisions, submissively accepting orders. As a corollary, a person in authority may become authoritarian and demand blind obedience at the expense of individual liberty – "Do it because I say so."

Doctors are "leaders" because they usually belong to the higher economic and social classes and are perceived to be highly intelligent and "hold the key to life and death". Within the health team the consultant is the doctor in highest authority. It is rare for a resident to question a consultant's order or to openly disagree with him. Physicians are also benevolent father figures. As such, they are to be respected and obeyed. Patients are usually submissive and inhibited in participating in their own care. Rather than question or refuse a proposed diagnostic test or treatment, a patient either agrees without understanding, asks another person for the explanation, or sometimes goes to another physician. A doctor, on the other hand, may not take the time or effort to obtain informed consent. He sees it as unnecessary; his judgment is to be respected and his orders obeyed. He believes that he is in the position of knowing what is best.

3.5 Utang na Loob

Utang na loob (internal debt of gratitude) is reciprocity or recognition of a debt of gratitude or honor that imposes corresponding obligations and behavior expectations. More than a simple financial debt which can be repaid, an *utang na loob* is an internalized debt which can never be fully repaid (Miranda, 1989, p. 52; Gorospe, 1988, p. 32). It demands that the recipient of a good act or deed behave generously towards his benefactors as long as he lives (Miranda-Feliciano, 1981, p. 70). It defines his integrity as a person in the context of social relationships (Miranda-Feliciano, 1981, p. 70).

Within the family it means that the children are expected to provide for their parents in their old age since they owe their life and entire education to their parents. This extends to older siblings who have helped in rearing and educating the younger sibling (Andres, 1981, p. 31). It is common to see a worker sharing his earnings to educate a younger sibling in order to repay his own debt to family. A child who does not pay his debt of gratitude, who fails to provide for his parents in their old age, is *walang utang na loob* – an ingrate (Andres, 1981, p. 114).

Outside the family, *utang na loob* is established after a gift is given. The gift may be material or otherwise, such as assistance in getting a job, admission to an institution, etc. Accepting the gift means the receiver will respond to some future demands of the giver, otherwise the gift should be rejected. Indebtedness may alternate back and forth between original giver and receiver until both become equally indebted and *utang na loob* partners. Individuals who are *walang utang na loob* risk being avoided or being avenged against, and may forfeit the security of the social group. Sometimes *utang na loob* is used to one's advantage. Alliances are made with influential friends or power figures for future use. A lowly person might campaign for a politician, then get his reward with a job after the election. A politician may stand as sponsor for a person of lower status, and then expect loyalty in his next campaign. When the *utang na loob* relationship involves a person of higher social status, repayment to the higher-status partner is never considered complete; the obligation remains always with the lower status partner.

Utang na loob affects the allocation of scarce resources, triage policies and implementation, and health professional relationships. The only available ICU bed is given to the person to whom one has an *utang na loob*. A residency position is given to the son of an *utang na loob* partner. A doctor will prescribe the product of the drug company that sponsored his latest trip because of *utang na loob*.

3.6 Bahala na

When apparently helpless in a difficult situation, the Filipino assumes a *bahala na* (come what may) attitude (Miranda-Feliciano, 1990, p. 14). It is a form of fatalistic resignation to events by withdrawal or shrinking from responsibility (Gorospe, 1988, p. 30), or just doing what needs to be done, when the prospect of success is bleak, with a daredevil acceptance or a defiant bravado (Miranda-Feliciano, 1990, p. 17). This has bred a sense of resignation (To Thin Ahn, 1994, p. 54), endurance (Gorospe, 1988, p. 60), absence of initiative and discipline (Gorospe, 1998, p. 60) as well as resiliency (Gorospe, 1997, p. 242).

Promotive and preventive health care is hampered when the Filipino does not take care of himself and takes unnecessary risks – continues to smoke, overeat or not exercise. Family planning measures have been unsuccessful partly because of *bahala na*. The Filipino couple "accepts" as many children as God gives them and trusts that God will provide. A patient agrees to proposed surgery, rarely asking explanations or seeking to understand consequences. When critically ill, despite prohibitive health care costs, the family instructs health care providers to do everything rather than stop everything, and leave the financial burden to resolve itself, accepting the misfortune as an act of fate or God.

4. PERSON ORIENTATION

The Filipino views his person holistically. He rarely isolates actions from person, parts (whether these be organ/cell/molecule/gene) from whole, mental from physical, disease from diseased. All these are viewed together as interacting with each other, forming the person.

Everything is taken on a personal basis. A child is criticized as being naughty and not as performing a naughty act. Even as an adult, when his work is criticized, the Filipino feels his person is criticized. A good person will not perform a bad act. A bad action is taken only by a bad person.

Disagreement is personal and a grudge is held long after the argument is resolved. Issues do not matter as much as personal allegiances. In politics the Filipino votes for the person, not for what he stands for.

Persons as a concrete presence takes precedence over abstract impersonal issues and ideas. Duty, responsibility and loyalty to a person far outweigh that to an impersonal institution. Secretaries often go with their boss when he changes companies. Those in authority tend to recommend or appoint persons they know rather than persons who are qualified.

The Filipino will repay his *utang na loob* to another person, but rarely recognizes a similar obligation to an institution that developed him. He views government or institution as having obligations to him rather than him having obligations to it. He willingly takes all the benefits he can get, using company supplies for personal purposes, or running personal errands on company time. Failing to do so is "missing out". An employee will attend first to a relative, friend or former mentor before doing his duties to his company, even if this violates company rules. Name-dropping is commonly practiced to facilitate process.

Similarly the health care provider recognizes his duties to his patient, but unless he is an administrator, he seldom recognizes his responsibility to the hospital he works in. He will do his best for his patient but has no qualms about doing less than his best for the hospital or gaining in an unmerited fashion from a drug company.

5. HEALTH CARE IS MULTIDIMENSIONAL

There are rich and poor Filipinos. There are illnesses of the rich (hypertension, coronaries, strokes, cancer), illnesses of the poor (diarrhea, typhoid fever, pneumonia, tuberculosis), and health care for the rich (Gorospe, 1997, p. 76). The society matron goes to the specialist in a tertiary hospital for a headache or hypertension, is confined in an air-conditioned room with a private nurse, gets a battery of sophisticated tests and a rigmarole of referrals, and is prescribed expensive imported name-brand medicines. The poor urban squatter or rural peasant usually waits until a disease is severe or terminal before she seeks health care. After a massive bleed, she is brought to the charity ward of some hospital or the local health center, the local *albularyo* or the shaman, whichever is available. Free treatment in crowded hospitals, herbal and spiritual healing, and being subjects for research and training are the alternatives forced on the economically depressed.

There is science, religion, superstition, and tradition (Buenavista-Tungpalan, 1997, p. 32). God and religion play dominant roles. The Filipino likes religious ceremony and favors joining processions, visiting shrines and participating in novenas. The patient is often seen clutching a rosary when entering the hospital for admission. Her family prays for her and asks others for prayers. She is surrounded by holy pictures, blessed objects, handkerchiefs which have touched relics. The patient seeks God's help in being healed and, even when accepting an illness as terminal, prays for a miracle.

Superstitions also affect behavior. Patients prefer not to be admitted before sundown or to Room 13. When healing is delayed, faith healers, fortune-tellers, quack doctors and witch doctors are called upon. An important reason why the peasant healer is popular is because he is emotionally and physically immersed in his healing technique, unlike many doctors who are more rational and emotionally detached (Miranda, 1994, p. 113; Gorospe, 1997, p. 77).

Traditional concepts of *pito-pito* (7-7), *usug* (bad wind), *lihi* (conception), and *hiyang* (suitability) remain common beliefs. Boholano mothers attribute cough primarily and most importantly to *piang* (fracture or dislocation) and therefore prefer to consult the *manghihilot* (local healer) rather than the doctor (Tallo, 1994).

In one and the same illness, the Filipino will see the doctor for the scientific diagnosis and prescription, use traditional concoctions from her family or local *albularyo*, have a statue at her bedside, and ask the priest for his intercession. These are not viewed as disparate or contradictory but as complementary and natural.

6. ISSUES

I must observe again that culture has good and bad effects. This essay is not meant to be judgmental. It is only meant to be sensitive to the effects of culture.

6.1 Autonomy

The Western value of the dignity of human person treats him as an autonomous, independent, rational human being, accepted as an equal regardless of age, fortune or social position. This allows him to be free to behave spontaneously, say what he thinks, to do what he chooses, with minimum social pressure. He is not overly concerned about the comments of others and is able to create his own life, enjoy his own uniqueness, listen to his own wants and desires, trust his reactions and feelings and be the "self which one truly is" (To Thin Ahn, 1994, p. 7). This encourages the practice of right to information; real respect for the privacy of each individual; confidentiality; and free and informed consent. In addition, this individualist/privatist attitude of "my mind, my body, my totality as individual person" has strong implications in decisions regarding abortion, transplantation and experimentation. The Western person makes his own decision.

Our Eastern tradition of social corpus through close family ties and belongingness, and of seeking to preserve harmony, extend the principles of truth telling, privacy, confidentiality and free and informed consent to the family where hierarchy is respected and paternalism is practiced. A disease is rarely private, a decision is rarely made alone.

Recently we had a patient we suspected to have AIDS. In the two weeks that he was hospitalized, I never saw him alone. How was I to ask him about exposure? How was I to respect privacy when suggesting a test for AIDS, and how could he freely answer? Finally we had to ask him the sensitive questions in front of his mother. When I told him we would test for AIDS I assured him only he, his mother and I would know about the test, as I would personally get the blood and have it anonymously tested. Confidentiality would be preserved. The following day his sisters were all in the hospital room and before I could speak they asked for the results of the AIDS test. They would together decide the next step to take.

6.2 Futile treatment and care for the terminally ill

From the Western point of view, a treatment which does not heal the disease is usually considered futile; technically and scientifically useless procedures are futile. For us, treatment might reduce discomfort, give time (there is a proper time to die, or time is needed for a relative abroad to come home), or show compassion. Feeding has a symbolic value of caring (Buenavista-Tungpalan, 1997, p. 138). These measures are not futile even if they do not heal; they serve the family, even if not the patient herself.

The same analogy is true when faced with newborns with multiple anomalies. Parents cannot withhold all treatment but ask health care givers to "do everything". Every child is precious.

6.3 Health Professional Relationships

Since seeking harmony is in our culture, how does one correct a colleague who is incompetent or unethical? How does one deal with pharmaceutical products from companies that provide educational support, gifts and hospitality when writing a prescription? These are difficult situations the health care professional has to face. He has to act ethically and always keep the patient's best interest foremost in his mind.

6.4 Health Policy

To be meaningful, any health policy should seek the common good and reach the common man. Common good cannot be disregarded for the sake of family interest or personal relationships. In our country, 90% of the population is poor (Gorospe, 1997, p. 27), and is untouched by scientific medicine and health care. Health policies must address the poor. They cannot be excluded. We cannot harm them (nonmaleficence) but must instead help them (beneficence). Health measures must be viewed from their perspectives. To the poor couple who see children as gifts from God, as a future income source and an insurance for their old age, who have sex as recreation, who assume the *bahala na* attitude with regard to how they will provide their children with food, shelter and clothing, natural family planning is often meaningless.

Many times folk healers and traditional medicine are all the poor can afford. To reject these as inappropriate treatment or suppress or discard them as medically irrational is too simplistic, especially when they are affordable, credible, accessible and acceptable sources of health care for the community. We should realistically weigh the benefits against the harms so that we optimize them and so that scientific and traditional medicine can collaborate with and complement each other.

Health policy must educate the rich, teaching them that to exploit the poor and uneducated as continuous resources for research and training is wrong. "Using others as ends and never merely as means", "The end does not justify the means" are not just words but should be put into deeds. Their eyes should be opened to the reality that there exists a connection between their lifestyle and the poverty around them (Gorospe, 1997, p. 341). There must be a dialogue between policy makers and the population so that real needs are addressed and good cultural beliefs and practices are preserved.

7. RECOMMENDATIONS

We should all be sensitive to the characteristics of our culture. Accept that culture with a certain amount of humility and a sense of humor, be conscious of its relationship to bioethics and articulate it in a systematic form. We should join the journey towards a global ethics: integrate diversity into unity, and develop from the different cultures "a fundamental consensus of binding values, irrevocable standards and moral attitudes," (Gomez, 1997, p. 124). In addition, when in a particular situation bioethical principles come in conflict with the cultural givens, we should strive and pray that we learn to use prudence and patience to discern what is truly ethically correct and culturally acceptable in the Philippines.

University of Santo Tomas
Manila, Philippines

NOTE

1. For simplicity, I shall use "he" and "she" without necessarily alluding to gender. The Filipino and the health provider shall be a he and the patient a she. This is not to mean that he is always truly a he or she is always truly a she.

REFERENCES

Andres, T.D. (1981). *Understanding Filipino Values. A Management Approach*. Quezon City, Philippines: New Day Publishers.
Alora, A.T. (1993). *Casebook in Bioethics*. Manila: Southeast Asian Center for Bioethics.
Buenavista-Tungpalan L. (1997). *Promises to Keep Return to Nursing*. Quezon City: Soller Press and Publishing House.
Gomez, F.G. (1997). 'Journeying towards a global ethics.' In: *Bioethics, the Journey Continues*. Manila: UST Publishing House.
Gorospe, V. (1988). *Filipino Values Revisted*. Metro Manila: National Bookstore.
Gorospe, V. (1997). *Forming the Filipino Social Conscience*. Makati City: Bookmark Inc.
Miranda, D. M. (1989). *Loob. The Filipino Within*. Manila: Divine Word Publications.
Miranda, D.M. (1994). *Pagkamakabuhay on the Side of Life*. Manila: Society of the Divine Word, Logos Publications, Inc.
Miranda-Feliciano, E. (1990). *Filipino Values and Our Christian Faith*. Mandaluyong Metro Manila: MF Literature Inc.
Tallo, V. (1994). *"Piang" "panahotor", the moon. The folk etiology of cough among Boholano mothers*. Manila: PSMID Convention.
To Thin Ahn. (1975/1994). *Eastern and Western Cultural Values Conflict or Harmony?* Manila, Philpines: East Asian Pastoral Institute.

CHAPTER 8

V. MANICKAVEL

LIVING IN SEPARATE AND UNEQUAL WORLDS: A STUDY IN THE APPLICATION OF BIOETHICS

1. THE LOCAL PROBLEM OF THE UNFORTUNATE AND LESS FORTUNATE*

Historically every community or nation has had its own way of solving the problem of the unfortunate in their society by arriving at a balance between the unfortunate and the fortunate or trying to contain the problem to a tolerable extent. Before the global era (B.G.E.), nations survived longer when a balance existed between the fortunate and less fortunate but less long when they tried to contain the situation without any ethical solutions. This latter situation of containing the problem without ethical solutions has resulted in revolutions and changes of ideology, which also tried to strike a balance between the less fortunate and more fortunate.[1]

At times the balance was struck by individuals who provided the necessary assistance. Disadvantages in health care were addressed by health care providers nursing and caring for the less fortunate; disadvantages in knowledge were addressed by those with knowledge giving or sharing their wisdom with the less fortunate; unfortunate victims of law and order were protected by the lawgivers. In all these cases the hierarchy and disparity between the fortunate and unfortunate was accepted, yet solutions were worked out and applied to maintain harmony.

2. THE GLOBAL PROBLEM OF THE UNFORTUNATE AND LESS FORTUNATE

The solutions to the problems of the unfortunate and disadvantaged were easily achieved with local involvement, and were mostly based on local culture values. In this way the situation became palatable to the culture and value of the local particular community. Deliberations on the solutions sometimes took longer if there were conflicts of interest and different benefit groups involved, but solutions were usually reached. Most of the time, the local moral values were the governing forces behind those immediate result-based solutions. For these local problems, the local moral values were able to give solutions in different historical periods and in different places.

R.-Z. Qiu (Ed.), Bioethics: Asian Perspectives, 83-89.
© *2003 Kluwer Academic Publishers. Printed in the Netherlands.*

   The local problem of the unfortunate or less fortunate now has become an international issue as national barriers are being removed in the process of globalization. I am not arguing in this essay that the polarization of fortunate and unfortunate had been widened and hastened due to globalization. Rather, I would like to argue that the local problem of the unfortunate or less fortunate had become an international issue in the globalization process, as now whole nations have become unfortunate or less fortunate relative to other nations. Let us focus on how the universality of ethical values and principles is being denied, concealed, or ignored in favor of other interests in the after-globalization era (A.G.E.).

   The modern economic growth of a nation depends on shifting to a market economy based on manufacturing and marketing. Globalization enhances and encourages such economic growth. One of the trends in modern times due to globalization is the separation of the manufacturing and marketing sectors. Thus, many nations have become involved only in manufacturing for international consumption and others are involved in marketing. Now, some nations are engaged only in production but not in consumption, and in many cases they become producers on the condition that they will not consume directly what they produce. The capital needed in many of these manufacturing sectors is prohibitive for the poor Southern nations, so usually the capital comes from countries that have excess wealth and significant marketing capacity and who may or may not have the need for the product. In many of the Southern countries, we now see specific areas or even towns that are created for export manufacturing. These export-manufacturing zones have no local taxes and are becoming the norm in many Southern countries. These countries offer many incentives, in the form of tax exemptions, tax holidays, and tax grants to attract capital from developed Northern nations. The national government may argue that such concessions are necessary in order to increase the number of jobs, reduce unemployment and boost money circulation. However, these tax concessions are not normally given to local manufacturers who manufacture and market to local people. This preferential treatment of multinational companies has created a reverse situation in which the entrepreneurs occupy the command seat. However, in B.G.E., commercial companies used to stand on the other side, and nations were in the controlling position of allowing or not allowing a business venture in their nations. Now the entrepreneurs are in the demanding position, and in order not to lose the entrepreneurs' investments, competitive nations give out more concessions, which sometimes lead to ethical tensions. Often they ignore ethical principles or values, so that "progress" can proceed.

3. FROM ECONOMICS TO ETHICS

Ethics is, among other things, concerned with truth-telling and honesty. One of the four principles of Georgetown's mantra of bioethics, informed consent, is based on truth-telling and information-sharing. Bioethicists are always concerned with dissemination of information without distortion and the assimilation of knowledge to make the right decision with informed consent. There is a large body of literature discussing the competency of consent-givers and their ability to comprehend and to make the right independent decision. Bioethically, these exercises are found to be important in the execution of research or clinical procedures. Furthermore, informed consent is generally essential in transcultural, transnational studies or research projects.

As the bioethics discipline per se is developing at a fast rate in A.G.E., a reverse argument is beginning to be heard in the nations where economic stimulation is sought. The argument is that economic stimulation should not be controlled by the application of stringent ethical principles. Under these circumstances, projects or studies which were, for ethical reasons, not able to be conducted in other countries (usually the country of origin of the donor company/entrepreneur), get transferred to countries where ethical considerations are not such a priority. In most of these situations, locals were allowed to dominate, and the universal governance of ethical values were undermined. This has led to circumstances in which transcendent ethical values, claiming universality, are being replaced by local moral values (Manickavel, 1997).

Now we have a situation like this: A running train is stopped in a station, but there are no signs of any transactions in the stations it formerly passed through. This may be for either of two reasons: first, the train did not stop in those passing stations because those stations were small; or second, the train did stop but the passengers did not want to get on. In this analogy, I am comparing bioethics to the train and the small stations as the nations in the development race. Whether the train did not want to stop at those small stations or the people in those small stations did not want to get into the train, the reality is that bioethics is being missed in these countries.

4. A CASE STUDY

Let us look at a case which has an international effect. This case pertains to medicine, the subjugation of women, and of course, poverty. Recently in India, the press and some women's organizations publicized an ongoing research study with a particular drug by the name of Quinacrine (Pollack & Carignon, 1993; Rajalakshmi, 1996). This drug was an old malarial drug which has now resurfaced for the purpose of a procedure called "Q-sterilization". This drug is administered as a tablet implanted in the uterus to sterilize women. Thus, it becomes an invasive medical procedure; only knowledgeable medical personnel can insert the tablet. Usually the physicians conduct the implantation procedure, and in India several were involved in this project. This large-scale study was financed by an organization in the United States. The money, tablets and implanting devices all came from the donor agency. It was promoted as a painless, non-surgical, effective sterilization procedure. The drug had been used before for this purpose in Asian countries like Korea and Vietnam, and other countries like Brazil and Mexico. The Indian government did not prevent the study from being conducted.

This tablet, when inserted in the fallopian tubes, causes inflammation and scar tissue formation within the tubes. As a result, the fallopian tubes become blocked. This drug was not recommended for sterilization by the WHO, as this blockage of the fallopian tubes could result in ectopic pregnancies and was also known to cause cancer in animal studies. Further, the WHO severely criticized this study for not collecting enough experimental data and because the preliminary studies had demonstrated harmful side effects. Previous studies in other countries received severe criticism due to the pain and suffering it had caused in women with several severe side effects. Neither the United States nor other Northern countries had approved this particular drug study for this purpose in their countries.

According to the Indian journal *Frontline*, this research project has been promoted and financed by two organizations in the United States (Menon, 1997) whom some have even alleged are supported by foundations allegedly held to have racist goals. .

> [T]he organisations ... are funded by anti-immigration groups .... For these groups, one way of preventing the burgeoning Third Word populations from swamping the US is to promote mechanisms of birth control that are both permanent and irreversible (Menon, 1997, p. 76).

As the author of the *Frontline* article describes it, although maternal health is given as the main motivation for the campaign, the financing organizations in the United States "also express their concern for the security threats posed by population growth in the developing world" (Menon, 197, p. 76).

Now, in this particular case, the local physicians who were responsible for this research conducted the study because of the host country's lack of requirements for strict adherence to bioethical principles. Meanwhile, the sponsoring agency could not carry out this study in its own country due to strict bioethical requirements.

These kinds of studies are usually conducted in countries where high levels of poverty and illiteracy prevail and where little professional sensitivity to bioethics may exist. Professionals from the donor country are also culpable in not following the regulations which are norms in their country of origin. The host countries, which allowed Quinacrine sterilization procedures on their women, made the argument that to ease the population pressure, immediate new means to reduce the number of births had to be explored. Thus, the host countries were not willing to adhere in this case to the principles of bioethics like beneficence, nonmalaficience, autonomy (individual/family/communal), truth-telling and justice.

Should not the sponsoring nations in this type of transnational research have an ethical responsibility? When a research activity from one nation gets transferred to a different country, should not the ethical regulation governing the research project also apply? If there is no sensitivity to ethical principles in the host countries, should the donor countries educate the host countries on these principles? Should there be two ethical standards? If so, why? Are universal bioethical principles not applied in the host countries because of double standards? To adopt the train metaphor once again, in this case the train did not want to stop and "deliver bioethics", whose delivery might interfere in the harvest of other benefits. So, the lack of ethical sensitivity in the host country and the different standards of application of ethical principles by the donor country contribute to the situation of unethical practice in the case illustrated.

Often decisions in bioethics are made on the basis of utilitarianism. However, in this essay and elsewhere (Manickavel, 1997; 1999), I have argued that the decisions in these areas are ethical only when they have more than merely immediate benefits. Recently, a quasi-utilitarian argument (Cooley, 2000) was proposed for the justification of Quinacrine sterilization on the basis of autonomy and freedom of choice. But, in this particular case, no real discussion about informed consent was conducted. For example, participants were not aware of the possible political motivations of the sponsoring agency. Selective disclosure about this and other facts certainly compromises the participants' ability to give informed consent.   In addition, though I am not critiquing all of Cooley's paper, it seems that there is the potential for justifying almost any research on quasi-utilitarian grounds. For example, pharmaceutical companies may use this sort of argument (i.e., the immediate possible benefits of a particular study) as a general justification for using drugs in some nations that have been banned in others.

Both the sponsors and hosts in cases like this may accrue individual, immediate and short-term benefits, such as individual fame, financial compensation, subsidized foreign travel, etc. Perhaps understandably, the former do not want to bother with bioethics and the latter do not want to be responsible for applying it. But the question should be raised whether it is wrong to aim at individual, immediate and short-term benefits. For example, in the Quinacrine case, sterilization and reduced births were considered to be more important than the permanent damage some women might experience as a result of the procedure. But it is important to remember that in this study, the subjects were women who historically have been subjugated, whose voice is traditionally not heard, and whose role in society is often undermined. In addition, only poor and rural women were used in this study. High- and middle-income, urban, educated women did not participate in this 'experimental' trial. Participants did not learn all the possible effects of the drug, and the proper informed consent was not obtained. It is not obvious that individual, immediate and short-term benefits are the only important moral considerations.

Now, let us look at a different scenario which quite often occurs. In most Southern countries, there is an acute shortage of electricity. In one of those countries the government wanted to alleviate this problem by building a very large hydro-electric dam. No local companies were able to undertake a project of such magnitude. Naturally the government wanted multi-national companies to bid on it, so to woo these companies, the government relaxed existing rules.

The local government was convinced about the advisability of the dam on the basis of immediate and short-term benefits such as the amount of power generated by the power project and the immediate revenue the sale of excess power could generate. Furthermore, the government was also satisfied by the large land area that would benefit from irrigation due to the dam. All these benefits were based on immediate, pure utilitarian values. The number of hectares which would be lost due to the huge catchment area, the number of biospecies which would be lost due to destruction of their habitat, and the number of people to be displaced due to the dam were ignored. The benefits to individual industries from the power project were not normally made available to the public. Most importantly, information on the effect of the dam in that ecosphere and the fate of the river were not usually studied or were ignored. The government normally neglected to analyze these other long-term, permanent implications and tried to apply alternative small intermediate technology to solve the problem. The long-term non-partisan or multi-national environmental assessment study is not usually conducted in Southern countries as it is in Northern countries. Multi-national companies normally use different standards of ethical evaluation depending on local politics and sometimes in the name of local culture. Local governments are not very particular about studying the long-term impact of the project, arguing that such a study is not warranted and that it will slow down the process. They also argue that the projected immediate and short-term benefits alone are necessary for progress, and all other benefits will impede progress. In reality, the immediate and short-term benefits are beneficial only for the local government in terms of opportunity to stay in power longer.

Recently a study has shown that along a cross-country highway project in Africa, the increased movement of transnational trucks increased prostitution in local communities. As a result, a high percentage of truckers and prostitutes are infected with HIV. The short burst of economic activity has in the long run placed a heavy burden on the health care of those communities (Platt, 1996).

The so-called local ethical reviews of the projects and studies are often biased and favor the views of the ruling party of the nation. In the Quinacrine study, the professional who conducted the research was also a member of the ruling party, and ignored the ethical reviews of the bureaucratic agency which conducted the project assessment. As a result of the press exposure, another Indian government agency has declared that the Q-sterilization procedure is illegal in India.

The greater weight put on the immediate benefits by the host country and the double standards adopted by the donor country contribute to the repeated global problems of environmental degradation and exploitation of weaker sections of the community. Less fortunate countries are becoming hunting grounds, and when the hunt is over, hunters from the fortunate countries leave for another hunting ground, where they are given all encouragement for another hunt in a virgin territory. In the long run the less fortune are becoming no better off. In this global exploitation, boundaries will come to exist where bioethics is applied. The repeated rape of environment and exploitation of weaker sections in Southern countries illustrates the lack of bioethical principles in action. This is due to the double standards adopted by Northern countries and the reluctance of Southern countries to follow bioethics, which is based on transcendent universal values.

Wherever a technology or knowledge or a business enterprise move to a virgin territory it is important that it also transfers the ethics which governs the technology. It is obvious that in a small village airport, a jumbo jet cannot land. But with modification and a strong long suitable runway, it can receive the jet for smooth landing and deliver the benefits to that community. In the same manner, the transfer of ethics along with new knowledge and technology will help the jet serve and benefit a new community. As an example, the initially unethical practice of kidney transplantation in India, which received global criticism, has now become accepted, along with the universal ethics of kidney transplantation. Thus the exploitation of the weaker members of the community has been averted.[2]

Ethical principles based on transcendent values always help humanity in providing justice to individuals, families, communities and nations. In all circumstances, the transcendence of justice is weighed against the immediate results or benefits. International laws at times become ineffective, but justice delivered with the responsibility for others on the basis of love is always effective.

5. CONCLUSION

In this essay it has been argued that ethics is not a luxury and should not be seen as a commodity only rich people and nations can afford. Further, it has been shown that it is a myth to consider ethics a stumbling block for progress. This second notion is a growing argument in many of the nations who are under the illusion of catching up in the race with developed nations. In that haste, Southern countries may feel that bioethical principles are unnecessary and Northern countries also seem to adopt different standards of acceptance when conducting research in foreign countries.

Kathmandu University
Bharatpur, Chiwan, Nepal

NOTES

* Part of this paper was presented at the Asian Bioethics Seminar, 9-10 November 1998, Nihon University, Toyko, Japan

1. Even though I have used the word "balance", it does not mean balance in the true sense of equal number. It is the balance struck where fortunate and the less fortunate were able to live together without much tension and harmony was maintained.

2. However, recently it has been reported that the exploitation is still continuing on but in a clandestine manner because of the initial lack of bioethics awareness and the already established business links (Chaudhury & Rajalakshmi, 1998; Mennon, 2002).

REFERENCES

Chaudhuri, K. & Rajalakshmi, T.K. (1998). 'Kidneys and crimes,' *Frontline Magazine*, 15(12), 43-46.

Cooley, D. (2000). 'Good enough for the Third World,' *Journal of .Medicine and Philosophy*, 25(4), 427-450.

Manickavel, V. (1999). 'Rich and poor; A paradigm of mutual co-existence?' Presented at the Second Asian Bioethics Seminar, November 1999, Nihom University, Tokyo, Japan.

Manickavel.V. (1997). 'Analysis of values on the basis of transcendency.' Presented at UNESCO Asian Bioethics Conference, Kobe, Japan.

Menon, Parvathi (1997). 'Questions of ethics and safety : The controversy over the Q-method of sterilisation.,' *Frontline: India's National Magazine* (May 2), 75-78.

Menon, P. (2002). 'Kidneys still for sale,' *Frontline: India's National Magazine*, 19(2), 33-35.

Platt, A.E. (1996). 'Confronting infectious diseases,' *State of the World 1996*, 114-132.

Pollack, A.E. & Carignan, C.S. (1993). 'The use of Quinacrine pellets for non-surgical female sterilization,' *Reproductive Health Issues*, 2, 119-122.

CHAPTER 9

UN-JONG PAK

BIOETHICS AND COMMUNICATIVE ETHICS

1. INTRODUCTION

This paper concerns the newly raised questions of medical ethics in the application of high technologies, including biotechnology. I address the limitations in applying the principle-based argumentation that has been popular thus far in bioethical discussions. I also suggest that we shift the discussion of medical ethics from the level of the ethics of institutions or the ethics of convention to the level of so-called "post-conventional ethics", the communicative ethics. In encountering the threat to life and the environment in the contemporary era of science and technology, justice is not enough. We need to reconstruct bioethics (including medical ethics) based on mutual respect and co-responsibility between equal moral subjects above the level of institutions and functional social systems.

For the discussion of the communicative ethics and ethics of care, I think the traditional thought on medical ethics in Asia, on the one hand, and the feminist perspective, on the other hand, will provide helpful instructions. However, I must confess that my thesis in this paper is incomplete and is still in process. My intention is to trigger further discussion by suggesting an initial idea.

2. SOCIAL AND ETHICAL CONTROVERSIES DUE TO THE DEVELOPMENT OF MODERN BIOMEDICAL TECHNOLOGY

Today almost all countries are exerting their best efforts in order to keep abreast of the rapid developments in modern technology and science. Biotechnology especially has shown a noticeable development under government support since it drew attention in the end of the 1970s. Even before we have begun to appreciate fully the concept of biotechnology, the term has already become commonly used, and has made a considerable impact on the development of modern technology in medical care.

Needless to say, modern biomedical technology should be protected and encouraged to develop as it is a product of free research of the human spirit and will eventually bring about considerable medical and economic benefits and ultimately promote the welfare of mankind. However, on the other hand, negative aspects of

R.-Z. Qiu (Ed.), Bioethics: Asian Perspectives, 91-103.
© *2003 Kluwer Academic Publishers. Printed in the Netherlands.*

science and technology should be regulated, since their abuse or misuse may cause irrecoverable harms to our society.

Modern biomedical technology is accused of being intricate, prohibitively expensive and potentially dangerous. For example, in determining the death of the human being, from the cardiological point of view, formerly we did not need much expert knowledge or special equipment. But today, when determining brain death, we require highly experienced professionals and a hospital equipped with expensive high-tech devices in order not to be entangled in an altercation over possible misjudgments among experts. Moreover, the question of brain death, combined with the further questions of organ transplantation as well as the concern for errors in judgement, often creates vague social anxiety about what could be the ultimate end of so-called "medical utilitarianism".

Due to the potential risk of modern technology in one way or another (Jonas, 1985, p. 3), there is a tendency to expand the scope of "negligence liability" in the fields of the environment, mass transportation, malpractice and the like (Pak, 1988, p. 57). With the development of biotechnology, the scope of liability in medical care is increasing. For instance, our legal system recognizes physician liability in cases where the physician did not inform the parents of the deformity of the fetus at such time that abortion was a possible alternative. As many are aware, nowadays the physician's duty of care has been retroactively applied to protect the human life that is not yet born or even conceived (cf. *Moores vs. Lucas*). Such an expanded duty of care is possible due to the development of technology, rather than due to our better appreciation of the integrity of the human being or the inviolability of human life. Furthermore, the unpredictability and inherent risk of scientific research has increased distrust of the scientist's freedom of research. This is reflected in the work of Robert Spaemann, a German legal philosopher, who argues that the skepticism of non-professionals towards risky technology should be recognized and not ignored (Apel *et al.*, 1984, p. 411).

Nowadays it seems that almost two-thirds of medical expenses are used for maintaining various medical equipment (Pak, 2000, p. 465). Ironically, the reality is that the costs for maintaining various equipment—which could not last more than several years—exceeds by far the medical fees for the treatment by the physician whose expertise could be acquired only through long-term training. This has made medical science more and more subject to modern technology, and increased its worrisome influence in changing the relationship between the physician and the patient, and even the notion of medical treatment (Lee, 1995, p. 130). In general, the physician's treatment is limited to a certain period, usually not long. However, once started, the intervention of expensive high-tech medical instruments is not likely to cease as long as the patient keeps suffering from the disease and can afford the treatment. Furthermore, the power of technology generates the myth that it can do anything, even what a physician cannot do. Such omnipotent force—the "Technological Imperative" (Purtilo, 1994, p. 439)—may save or destroy human life and increase the ratio of our medical expenses to GNP.

Second, the development of high technology is accelerating the "polarization" in medical treatment. About a decade ago, the World Health Organization (WHO) declared that it would achieve worldwide health welfare by the year 2000 (WHO & UNICEF). However, while developed countries are debating on the issues of euthanasia, organ implantation, genetic treatment and the like, developing countries are still having difficulties in obtaining basic medical care. Today, it is often heard that while the poor die due to the lack of basic medical treatment, the rich die of pathogenic disease. With the advance of sophisticated technology, such polarization will be aggravated both nationally and internationally, which may eventually increase social anxiety and deepen the gap between the poor and the rich (Olwney, 1994, p. 170).

Third, the business of medicinal care or organized medical services was also made possible more or less by the development of high technology. Considering the huge administration and management system of current general hospitals, previous days look happier and more humane in retrospect in that medical ethics was then regarded as based on a personal relationship between the patient and the physician. In the past twenty or thirty years, legal theories have been developed to resolve issues including "who has the initiative between the patient and the physician in medical treatment?" Now, however, we are facing new issues such as corporate ethics and organization ethics in medical treatment. With the acceleration of the business of medical services, the medical care system might exacerbate the problems of the isolation of patients—ultimately even the isolation of physicians!—and disrupted communication (Kim, 1986, p. 13).

Could corporate or institutional ethics also be considered in medical care? Regardless of their for-profit or non-profit status, medical care-related entities are distinguished from other entities since they address human life. While other enterprises consist of contractual relationships such as transactions or selection between equal parties, the medical care enterprise works on transactions between the patient and the physician, which involves unavoidably unequal relationships in terms of information and technology. Although we use different terminology for the patient and the physician, such as "the insured" and "the insurer" or "the employee" and "the consumer", the nature of the unequal relationship will not change. Therefore, we will need to develop new ethical standards to evaluate institutional behavior, and to avoid the extremist approach which contrasts the economic (or profit-oriented) perspective and the moral perspective and then forces a choice between financial gain and professional responsibility. The medical-care-related entities will need to prepare their own ethical standards under which they encourage the ethical behavior of individual members. Further, if necessary, specific legal standards should be established better to incorporate and enforce such ethical standards in society (Mariner, 1995, p. 238).

Fourth, with the development of biotechnology, there has been a tendency to overemphasize the genetic features in medical care. Along with the great achievements in genetics, there have been many reports that almost all diseases are connected with genes in a certain way. How should we respond to such overemphasis on the genetic factors? What impact will it have in our society in the long run? Such overemphasis may function to underestimate the gravity of the social and economic situations such as malnutrition, the unsettled medical care system, car accidents, drinking, smoking, drug abuse, pollution, violence, housing problems and much more. Further, we may eventually encounter a new ideology: so-called "genetic determinism", under which we convert almost all social issues to genetic issues. Regarding the question of genetic tests or examinations, some bioethicists suggest a "right to ignorance"—a right not to know more than they want to—as one of an individual's basic rights, which trumps the "right to know" of interested parties such as health-related authorities, spouses, employers and the like. As long as it is concerned with genetic conditions, if ignorance of their fate is the only means to allow them freedom, "a right to ignorance" should be protected under the Constitution.

In addition, genetic knowledge seems to be changing the concept of medical treatment from "therapy" to "understanding" and "prediction". Medical science was originally based upon "therapeutic medicine" rather than "predictive medicine". However, nowadays the advancements of genetics are moving medicine towards predictive medicine (for example, genetic analysis prior to pregnancy) rather than therapeutic medicine or creative medicine. Predictive medicine may be used to encourage a so-called "defensive treatment" in order to avoid the risk of malpractice suits. As previously mentioned, in one case a physician who did not inform the family of the genetic defect of the fetus in a timely manner so that the fetus could have been aborted was found by the court to be liable. If physicians too frequently conduct genetic tests in order to reduce the risk of malpractice suits, such frequent testing may impair the health of the patient and cause undue financial burden. In addition, some physicians may be over-conscious of genetic features, and implicitly recommend abortion even in cases where abortion is not necessary. Accordingly, in order to reduce the social costs of such defensive treatment, the court should not compel physicians to conduct all available diagnostic tests on a patient. Given that "defensive" medical science has a tendency to expand the scope of abnormality and disease, at this stage our legal system should work to limit the scope of such abnormality in medical diagnosis. In this regard, we should provide a forum for discussion to enhance the mutual understanding between legal scholars and the physician.

Legal scholars expect that we will need to have a certain form of legal protection concerning individual genetic information. In order to obtain social consensus on this issue, we will need to have a public discussion regarding whether an individual should be discriminated against by reason of his or her genetic features in social security, insurance, employment and the like. (In the future, an individual who does not have a dangerous genetic feature may request a decrease in his or her medical care insurance premium just as a car insurance premium may be decreased if a driver has no prior record of accidents.) There may be an increasing tendency to recommend abortion if the physician discovers a slight chance of disadvantage in genetic features (although there is no actual indication of deformity). If that were the case, Steven Hawking might not have been born and achieved so much in science. Such a tendency is certainly against the notion of "human dignity" under which we accept the disabled as a member of our society. And our fundamental social understanding and structure will eventually be impaired by such tendency.

With the great achievements in genetics, medical science is undergoing a transition period, where it can merely *predict* the possibility or risk of disease without providing a complete solution. At this point, in light of increasing genetic consultation, diagnosis, and treatment, there appears to be an urgent need for the "communicative model" of ethics, based upon mutual understanding and trust between physicians and patients (including their family) rather than on the mere provision of information or paternalistic intervention.

Fifth, the "utilitarian medical model" born by medical technology may have such a powerful impact that it could influence not only personal perceptions but also family relationships, socio-economic structures, and discourse on the quality of life at large. For instance, treatments using embryonic tissue may revolve around this model. As the number of diseases treated by transplanting the embryonic nerve tissues to adult brains (as in Parkinson's disease treatments) increases, utilitarians have even argued that as long as there is no alternative, embryonic organs may be used without any limits. Due to the natural advantage of fetal tissues or organs (for example, the fast growth of the cell and organ, easy adaptability of a transplanted organ to a new environment, etc.) in modern medical treatments, fetal cells or organs are often used to obtain basic biological and medical knowledge, and sometimes such transplantation is even conducted for cosmetic purposes.

In addition to organ transplantation, the use of fetal organs becomes one of the more complicated and serious social issues, since abortion can be performed from the economic as well as humanistic point of view. In this regard, several countries have recently prepared guidelines for the use of fetal cells and organs to the effect that donation of fetal organs and cells should not adversely affect the time and the manner of abortion (Pak, 2000, p. 390).

3. THE RELATIONSHIP BETWEEN PATIENT AND PHYSICIAN

Unlike in the past when most patients in the hospital were suffering from acute or epidemic diseases, nowadays, those with various chronic or incurable diseases dominate the hospital population and are the main beneficiaries of sophisticated medical technology. As the reader may be aware, such modern diseases cannot be easily cured within a short period of time. Often the physician can merely sustain the life of the patient via the use of advanced medical technology. Further, with the development in genetics, almost all diseases are found to be connected with certain genes, which will force medicine to change from its traditional approach to other approaches better suited to the new genetic knowledge. Among other implications of this is that consistent care and counseling may become a more important kind of medical care than emergency treatment. A possible development for 21st-century medical services would be the division of medicine into two types of medical care, namely, 'therapeutic' and 'counseling'. The former would merely provide traditional medical service with the help of high-tech medicine and the latter would provide comprehensive medical services in order to prevent possible disease and promote counseling for the patient's health.

Modern biomedical science has changed the concept of medical treatment and accordingly the decision-making model in medical ethics as well. Traditionally, medical ethics has been understood based upon the relationship between the patient and the physician. In other words, the traditional decision-making model consisted of a physician's explanation and recommendation on one hand, and a patient's autonomy to consent to or reject that treatment on the other hand. Therefore, the physician's duty of confidentiality, the patient's right of privacy, and a *bona fide* relationship between patients and physicians were all traditionally emphasized, while other factors were relatively less important in decision making. However, according to recent surveys, the factors considered in such decision making are found to be different from what we had thought. In reality, patients rely in the decision making process much more on communication with family or others than with the physician. Eventually, most decisions are made by the patient her/himself with the support of other family members. Especially in chronic disease, the patient does not consider the physician as the only expert to consult. He or she may also consider his or her own experience and pay more attention to his or her family's opinion. Further, it has been reported that the patient considers non-medical factors, such as extent of financial burden, quality of life, and family impact, as more important than medical concerns such as effectiveness of medical care, risk of operation, etc. (Reust, 1996, p. 44). To the physician, the disease may be just a subject of medical research, but to the patient, the disease is a part of his life. Until now, when discussing medical ethics, we have not considered the patient as a "social being" and paid enough attention to the non-medical factors. As a result, we have totally ignored the personal experience or ability of the patient and his or her family in the process of conducting medical treatment. We have not regarded it as a duty of the physician to make a reasonable explanation to patients and their families regarding the inherent risk or limits of the medical treatment, and to make a sincere effort to help them understand the medical situation. It has been mistakenly believed that patients are merely recipients of medical support and cannot provide any help to the medical practice. In fact, the attitude of valuing a patient's individual experience has been advocated as an important facet of traditional Asian biomedical science.

In some respects Koreans have taken the current attitudes of doctors for granted, who may not treat patients tenderly and who only treat disease without consideration of the cause or the prevention of the disease. Very few patients might have the good luck to be greeted and sent off by doctors or other medical staff. How many people are now leaving the hospital without knowing their real condition or even daring to inquire about it? Due to the lack of communication like this, patients and their families, once so compliant and obedient to doctors, have changed their attitudes abruptly to take action against doctors.

However, the ancestors of Oriental medical care were not like that at all. They treated 'patients' rather than 'disease'. For example, the Korean medical scientist and doctor Jema Lee (1837-1900) named a disease not after its symptoms, but after the patient. They worked on preventive medical care and helped patients to develop their own measures to maintain their health, and they believed that the treatment could not be fully effective unless patients and doctors had confidence in each other.

Their study was not confined only to the affected area. Rather, they considered the cause of disease in the more comprehensive context of the interrelationship of mind and body, living environment, nurture, dwelling situation and the like. They understood their job of treating disease in the context of constructing a sound and just society. They developed medical science based on life patterns that are friendly to the environment and suitable to natural principles. Heo Jun (1546-1615) thought it important that the general knowledge of medical science should be disseminated as widely as possible and thus come into popular use. This approach of medical ancestors to disease and health reminds us of the core of the medical ethics: the *bona fide* relationship and the appropriate communication between patients and doctors.

A big gap still exists between the imported Western medical system, which centers on the symptoms and information provided by the professionals, and the general understanding of medical services which is expected to be centered on persons and the *bona fide* relationship of patient and professionals. The concepts of the patient's right of autonomy and the physician's duty of explanation were first introduced through the Western legal system. In Korea, the concept of "informed consent" has not been fully appreciated so far. Such a Western approach does not seem to be of much help in Korea, partly because of the Korean traditional views of disease and life or death, and partly because of the reality of our medical service system. In reality, even the physician himself can hardly set a limit on what constitutes a reasonable explanation—i.e., how far and to what extent he should explain a situation to a patient—considering the tremendously complicated and diverse modern biomedical technology. It appears that physicians are not content with the court's decisions on tort liabilities or malpractice liabilities in medical cases based on this criterion of the duty of explanation. Moreover, some physicians consider the duty of explanation as sometimes hindering their medical treatment. However, it should be noted that the use of informed consent, obtained when the physician has given reasonable explanation and the patient has a full appreciation of the relevant factors, is not intended to ensnare the physician in trivial cases. Instead, it highlights the fact that appropriate communication between patients and physicians should be an essential subject of the medical ethics.

There are not a few cases where patients are inclined to fail to behave with autonomy and responsibility despite the physician's full explanation of diverse options and alternatives. Instead, they take a passive attitude and follow the physician's decision at the crucial moment. Some Asians still take the matter of life and death as their destiny and do not feel comfortable with the idea of prolonging their lives against nature ("Omakase" patient in Japan) (Ishiwata *et al.*, 1994, p. 62). In this regard, there is an old saying that when there is no disease to suffer, it will be the day to die. When cancer patients were interviewed, half of them said that they did not wish to be told of the truth by the physician. Further, given the reality that many people formally sign a form to the effect that they will not take any legal actions against the physician's malpractice prior to the operation, the imported Western-style medical ethics is likely to remain only a nebulous discourse.

The concepts of the patient's right of autonomy and the physician's duty of explanation were first introduced through the Western legal system. In Korea, the concept of "informed consent" has not been fully appreciated so far. Such a Western approach does not seem to be of much help in Korea, partly because of the Korean traditional views of disease and life or death, and partly because of the reality of our medical service system. In reality, even the physician himself can hardly set a limit on "the degree of the sincerity of the reasonable explanation"—i.e., how far and to what extent the physician should explain to the patient—considering such tremendously complicated and various modern biomedical technologies. Thus it looks like that they are discontented with the court's decision-making on tort liabilities or malpractice liabilities in medical cases based on this criterion of the duty of explanation. Moreover, some physicians consider the duty of explanation as frustrating their medical treatment. However, it should be noted that the patient's informed consent with a full appreciation of the relevant factors upon the physician's reasonable explanations thereof is not intended to snare the physician in trivial liabilities. Instead, it is worthy of note that appropriate communication between patients and physicians should be an essential subject of medical ethics.

There are not a few cases where despite the physician's full explanation of alternative options, patients are inclined to fail to act with autonomy and responsibility, only taking a passive attitude to follow the physician's decision at the crucial moment. Some Asians still take the matter of life and death as their destiny and do not feel comfortable with the idea that they will prolong their life against nature (such as "Omakase" patient in Japan). In this regard, there is an old saying that when there is no disease, that is the day to die. When the cancer patients were interviewed, half of them said that they did not wish to be told of the truth by the physician. Further, given the reality that many people formally sign a form of undertaking to the effect that they will not take any legal actions against the physician's malpractice prior to the operation, the imported Western style medical ethics is likely to remain only as a nebulous discourse.

The concept of the "informed consent" has been developed since 1970 in Western countries including the U.S.A. as one of the means to fill the gap between the law and the medical practice concerning the physician's duty of explanation and the patient's right to consent. Some scholars are skeptical about the duty of explanation since there seem to be too many possible treatments to provide all the relevant information to the patient. Yet the concept of informed consent is worth notice as a concept in the process of developing in constant efforts to narrow the gap between the law and the medical practice.

On the other hand, physicians are more likely to be utilitarian or instrumentalist in considering primarily the disease of the patient rather than patient's rights or the impact on his family. Physicians are quite accustomed to the traditional interventionist paternalism and to treating patients like children who know nothing about medical science and who merely await physician's treatment. It would be more desirable to reject both the extreme normative approach and the medical exclusive approach. The former regards patients as isolated subjects with autonomy without considering them as they are, while the latter considers patients as objects expected to cooperate in procedures in accordance with the physician's intent. As high-tech medicine advances along with biotechnology, the issue of "communication" will become a primary concern, especially during the initial stages of a physician/patient encounter. Medical malpractice suits are often begun out of the rage of patients driven by a physician's failure to communicate in one way or another.

The concept of "informed consent" does not mean that one party enjoys a right and the other party merely bears an obligation. Rather, informed consent implies an interactive relationship in the process of decision-making. This is not an explanation in terms of the exchange of information. Rather, it is a concept for ensuring a partnership based on mutual understanding, which means, for example, asking "if the physician were in my situation, what would he do?" and *vice versa*. (Recent articles published in Western countries have discussed several methods of communication for "informed consent". These methods include a reasonable eye contact with the patient, a gentle attitude of listening and respect so as to induce the patient's questions and interest, a good tone of voice, professional behavior and look, etc. (Silverman, 1996, p. 228).) In order to avoid converting respect for human beings (in the sense of autonomy of the patient) into lack of personal interest in the life and death of others in society, the promotion of the autonomy of those in need is inevitably related to the responsibilities of physicians. In other words, this concept could be understood as an expression of care, concern, and social solidarity. If understood in this regard, several misunderstandings concerning "informed consent" might be dispelled (Meisel, 1996, p. 2522). First, in practice informed consent seems to be a matter of formality that merely requires a signature by the patient. Second, it becomes an another type of "Miranda" warning—it requires "officials" to inform the "suspect" of the right to remain silent and the right to counseling, etc. It is believed that the mere warning of the risks of treatment is enough.

Another potential problem with informed consent stems from the trend towards the medical "team" system, especially in general hospitals. In such situations, patients can often not distinguish between medical students, physicians, interns, etc. Assigning the task of informing the patient to an individual member of the team (and informing the patient of this fact) might lead to better communication and cooperation on the part of the patient. A notification system that requires the hospital to inform the patient of the role of the medical students in medical care should be introduced in Korea.

4. THE NECESSITY OF COMMUNICATIVE ETHICS

4.1. Traditional Discussion in Principle-Based Bioethics

Bioethics, as one form of applied ethics, is supposed to deduce ethical judgments from a series of principles. That is, it will apply the series of principles in actual cases to reach a conclusion. Today, there are four generally accepted basic principles in Western medical ethics: i) respect for autonomy; ii) nonmaleficence; iii) beneficence; and iv) justice.

4.2. Limits in the Principle-Based Approach

There are some advantages in approaching bioethics based upon these principles. Such an approach may remind one of the importance of cases that are likely to be ignored by deductive analysis. Since some believe that ethical judgments should be deduced from the structure of theories or general norms assumed beforehand, they do not pay much attention to practical cases in general. Given the unknown aspects of advanced biotechnology, case study (as opposed to the use of abstract theory) is very important. Since the collection, analysis and comparison of various cases is deemed to be more important than the deduction of answers from general moral principles, the rash application of these principles and the use of procrustean judgment may eventually aggravate suffering. In practice, the principles may contribute to reasonable judgment on ethical issues, oscillating between general norms and actual cases.

However, does it suffice merely to review bioethics based upon the principles? Can the principle-based argument be a plausible moral argument, grasping the sensitiveness and the importance of individual cases? Can it reveal the dimension where the questions of ethics and suffering are deeply related to the environment surrounding a particular individual, class, or gender? What matters in bioethics and medical ethics is accurately to understand the situation and to suggest a solution. There lies the reason why we could not get a satisfactory answer merely by appealing to theory or general principles in resolving bioethic issues. For instance, concerning artificial insemination, genetic testing or informed consent, the understanding of the context plays a very important role. These are the reasons why the issues of euthanasia or a physician's duty of confidentiality could be better resolved through the ethics committee rather than before the court. In certain aspects, it seems to be impossible to discover one general principle to apply to all cases, and it may be that this seeming impossibility is one of the major characteristics of bioethics (Sherwin, 1992, p. 42).

In order to understand the issues better, we should carefully review why the four principles discussed above cannot work well in the current system. For example, the "principle of respect for autonomy" is based on the idea of a society of freedom, equality, and distributive justice. However, in cases where one individual or one class is subordinate to another individual or class, a request for autonomy might be a cover for a selfish request. Further, "the principle of nonmaleficence" may provide a pretext for a physician's reluctance to take a more aggressive treatment for the patients. On the other hand, "the principle of beneficence" may be the ground for the physician's excessive treatment of patients under the name of pursuing science. Further, while seeming to present various virtues on its face, "the principle of justice" can, in fact, be abused to justify non-democratic distribution that has no consideration for different individual experiences, values and physical conditions (Tong, 1997, p. 244).

Now the concept of communicative ethics as well as the limitations in principle-oriented bioethics may again draw our attention. Today, it is not exaggerating to say that complaints from consumers of medical services are related to the amount of communication between physicians and patients. In this regard, the development of the new communicative model has emerged as one of the important subjects of bioethics.

4.3. Communicative Bioethics

We often experience in our daily lives that communication changes interpersonal dynamics. The dignity of a speaker might invite further discussion or challenge, or, on the contrary, block the natural flow of the discussion. This means that communicative behavior may incorporate power structures in practice. In this context, recently some bioethicists (noticeably feminists) have tried to re-establish bioethical theories based upon principles of communication. The enterprise of communicative ethics from the feminist perspective does not lie in merely enhancing the communication between patients and physicians. It instead aims at encouraging consensus regarding new legal systems and policies, and revealing the unequal power of parties in medical care and attempting to remedy it by establishing a new medical communicative model.

In medical ethics, the necessity of communicative ethics has to do with the changes in the concept of medical care described above. Nowadays, as various chronic and genetic diseases represent the majority of modern disease, the concepts of medical care, health, life and death have also undergone considerable changes. This means that between patient and physician, the importance of good communication, counseling and encouragement is increasing. Despite this, the importance of the physician's communicative ethics has been almost disregarded in medical education. Some medical students are even surprised at the fact that they need the patient's approval to proceed with a particular treatment.

If we wish to have a physician with good communication skills, ethics should be necessarily characterized as "considerate" and "communicative". Indeed, communication based upon mutual understanding and agreement is the essence of ethics, and the aforementioned patient's right of refusal in medical care or the concept of informed consent are deemed to be quite important in medical ethics.

Care-oriented ethics and communicative ethics are the most controversial subjects among feminist bioethics. However, these two fields are most attractive to and favored by scholars who are trying to reconstruct a new medical ethics through the reinterpretation of Asian traditional values and the perception of the community.

Feminists tend to expand the concept of circumstantial factors like "considerate" or "care" into the notions of generosity, social solidarity, mutual trust and the like (Udovicki, 1994, p. 49). They pay attention to the moral and emotional concerns that could bring greater moral efficiency, rather than merely arguing on the basis of abstract principles of justice and resulting rights. Such moral and emotional concerns have been considered in connection with the less self-centered and closed relationships such as friendship, parent-child relationships and the like. In such relationships, the requests of the "concrete other"—not the favor of the "generalized other"—are the focus of consideration. Even situations in which the principles of nonmaleficence or beneficence should be applied are associated with complicated interests including trust, generosity, cooperation, partnership, mutual respect. Features of such cases may go beyond objective and general concepts such as justice and non-interference. Some feminists have argued that this less self-centered relationship is the place where the "thick concept of morality" governs, and "general relationships" is the place where the "thin concept of morality" governs (Udovicki, 1994, p. 55).

There also has been an attempt to reinterpret Buddhism and Taoism and to focus bioethics on caring and relationships. Feminist philosopher Ok Hi Shin has introduced Wonhoy's (617-686) thought that "people can realize themselves only through relationships with others" as the "Buddhist model of care ethics", and tried to develop traditional Oriental ideas as a "relational ontology" (Shin, 1998, p. 20). Some feminists have taken the notion of "being unattached", or having an "absence of ego" from Buddhism and Taoism in claiming that humans should discover their genuine identity in the mutual dependency and relations of human beings (Kim, 1998, p. 37). Further, they seek a model of the "non-egoistic woman" who may respect the life activities of all creatures under the natural law, liberate them from prejudice or bias, and finally achieve the entire elimination of self and impartiality. The "non-egoistic woman" is one who has acquired identity as "a being that has undergone an incessant self-improvement process through relationships and obtained a high level of self-autonomy." In "the true relationship," the ego need not be bound to certain fixed substances like sexual identities, and finally can exist as a social and universal ego (Lee, 1998, p. 54).

It is natural to include various concerns such as relations to others, context or circumstances, and particularity into the discussion of ethics, because it is true that the human being is the so-called "relational being." The feminist's relation-oriented ethics has significant implications in its criticism of traditional ethics and its shift of the emphasis of morality to relationships. However, since humans are trying not only to have relationships but also to be separate from them at the same time, other considerations like universality and justice also must be seen as morally important. The problem with feminist theory lies in the possibility that it may impose sacrifice and concession on those with limited autonomy, especially in a hierarchical society where the imbalance of powers is dominant. For instance, it has been observed that under the relation-based Confucian ethics, ethical norms were adopted in the form of social norms where the ruled classes (including women) were unilaterally compelled to obey the ruling class.

It is true that natural compassion for others' pain is not merely to be recommended as the appropriate "female" moral attitude, but is rather a general norm for society. However, if there is no equality in a society, such a norm could be abused as an ideology of the ruling class. That is the reason why care ethics cannot comprehend all aspects of morality. It is on the basis of this point that I argue it is necessary that relational ethics should develop into rational communicative bioethics.

Relational feminist bioethics emphasizes that a physician should consider patients as "concrete others" and "generalized others" simultaneously. This is because, for example, while patients need not necessarily consider the impact of his father's death on the physician's thoughts concerning life and death, the reverse may not be true. Relational ethics reminds us that the moral subject of autonomy is not an atomistic, autonomous being existing exclusively in his or her own arena, but rather a related and autonomous being who can incorporate requests of others in his own individual norms. This approach criticizes the "informed consent" model for assuming a too-rational human being. Between patients and physicians, it is necessary to formulate an interpretive and more deliberate model, rather than a paternalistic or informative model.

Roughly speaking, the subjects of communicative ethics between patient and physician are related to the questions of overcoming power imbalances in health care, and of remedying the exhaustion of the consumers of medical services caused by the expansion of the medical market. For this reason, when discussing new ethical issues, we should consider the human being as a whole, incorporating both spirit and body, and try to introduce emotional concern or solidarity into the ethical decision making process rather than excessively depending upon reason and critical analysis. Ultimately this could lead us to achieve the true meaning of communication based upon mutual understanding (Smith, 1996, p. 184).

The legal system is not a product of abstract reasoning. Instead, it is a product of the meaning shared among people in a particular society. Communication in traditional medical care was nothing but a model of exchanging information or answering questions. This was fairly mechanical and physician-oriented in that the physician unilaterally listens to the patient and provides information concerning disease and its treatment. In such communicative relations, the physician normally tends to mention only one option among other possible alternatives. However, given the authority of the physician, the patient might take such an option as the sole and absolute solution. In other words, under the existing communicative system in medical care, only the professional has unilaterally decided what to mention, how to mention it, how long to continue the conversation or when to change the topic. Such unilateral communication in the form of questioning would never make reasonable information exchange possible, but merely confirms who has the stronger power in the end.

Ewha Woman's University
Seoul, Korea

REFERENCES

Oriental Classics:

Donguibokam, Neryang-puyn, Pungnyunsa, 1996.
Donguisusebowon, Uiwonron,Yeokangchulpansa, 1992.
Whangjenekyungsomun, Soohkwharon-pyun,77, Komunsa, 1971.

Modern Publications:

Apel, K.-O., Boehler, D. & Rebel, K. (Hrsg.) (1984). *Praktische Philosophie/Ethik.* Weinheim: Beltz.
Jonas, Hans (1985). 'Technik, Ethik und biogenetische Kunst: Betrachtungen zur neuen Schoepferrolle des Menschen.' In: Rainer Floehl (Hrsg.), *Genforschung-Fluch oder Segen?* (1-15). Muenchen: Schweitzer.
Ishiwata, Ryuji & Sakai Akio (1994). 'The physician-patient relationship and medical ethics in Japan,' *Cambridge Quarterly of Healthcare Ethics* 3(1), 60-66.
Kim, Jong Yul (1986). 'Corporate ethics and medical ethics,' *The Korean Hospital Association Magazine,* 7, 13-20.
Kim, Jung Hi (1998). 'Ontological research on biofeminism -- Consideration of new human model based upon the conscious in the Buddhism and Taoism.' Ph.D. Dissertation, Ewha Woman's University.
Lee, Boo Young (1995). 'Professional responsibility of the physician.' In: *Medical Ethics in Modern Society* (pp. 122-133). Asan Sahoebokchi Saupchedan.
Lee, Sook In (1998). 'Feminist interpretation on relational ethics in Confucianism.' In: *Feminist Ethics Theoryand Application.*
Mariner, Wendy K. (1995). 'Business vs. medical ethics: Conflicting standards for managed care,' *Journal of Law, Medicine and Ethics* 23, 238ff.
Meisel, Alan (1996). 'Legal and ethical myth about informed consent,' *Archives of Internal Medicine,* 156(22), 2522 ff.
Moores v. Lucas, 405, So. 2d1022 (Fla. 1981).
Olweny, Charles (1994). 'Bioethics in the developing countries--Ethics of scarcity and sacrifice,' *Journal of Medical Ethics,* 20, 170ff.
Pak, Seung Seo (1988). 'Medical ethics in legal text,' *The Korea Hospital Association Magazine,* 5, 57ff.
Pak, Un-Jong (2000). *Law and Bioethics in the Age of Biotechnology.* Seoul: Ewha University Press.
Purtilo (1994). 'Medicine in transition -- Emerging ethical issues in the global village', *Annals of the Academy of Medicine* (Singapore), 23(3), 439ff.
Reust, Carin E. (1996). 'Family involvement in medical decision making,' *Family Medicine* 28(10), 39-45.
Sherwin, Susan (1992). 'Feminist ethics and medical ethics.' In: Shogan, Debra (Ed.), *A Reader in Feminist Ethics* (pp. 42ff.). Ontario: Canadian Scholars' Press.
Shin, Ok Hi (1995). 'Traditional ideology in the Orient and the prospects of the Korean feminist philosophy.' In: *Han'guk Josongchuihak (The Korean Feminist Philosophy).* Hanwool Academy: Feminist Research Association.
Shin, Ok Hi (1998). 'Feminist ethics from the life of Korean women's viewpoint.' In: *Feminist Ethics -- Theory and Application,* Korean Association of Women's Studies, Fall Term Seminar, June 13, Data, 20ff.
Silverman, David R. (1996). 'Narrowing the gap between the rhetoric and the reality of medical ethics.' *Academic Medicine,* 71(3), 228ff.
Smith, Janet Farrell (1996). 'Communitive ethics in medicine: The physician-patient relationship.' In: Susan M. Wolf (Ed.), *Feminism & Bioethics: Beyond Reproduction* (184-215). Oxford: Oxford University Press.
Tong, Rosemarie (1997). *Feminist Approach to Bioethics.* Boulder: Westview Press.
Udovicki, Jasninka (1993). 'Justice and care in close relationship,' *Hypatia* 8(3), 48-57.
WHO & UNECEF (1987). International Conference on Primary Health Care, Alma Ata, USSR, September 6-11.

CHAPTER 10

LEONARDO D. DE CASTRO & ALLEN ANDREW A. ALVAREZ

SAKIT AND KARAMDAMAN: TOWARDS AUTHENTICITY IN FILIPINO CONCEPTS OF DISEASE AND ILLNESS

This paper aims to highlight the value of authenticity in responding to disease and illness. The argument is that authenticity must be added to utilitarian and pragmatic considerations as a criterion of the validity of healing responses to disease and illness. Authenticity may be achieved by ensuring the alignment of treatments and remedies with the social and cultural dimensions of concepts of disease and illness. The point is that diseases and illnesses are not merely biophysical phenomena. They form part of a matrix of values, traditions and beliefs that define a cultural identity.

To begin, this paper makes a distinction between two health-related concepts based on the meanings and uses of two words that are commonly used in Tagalog-based Filipino to refer to either disease or illness. These two words are *sakit* and *karamdaman*.[1] In ordinary discourse, *sakit* and *karamdaman* are often used interchangeably. However, there are important nuances in meaning that may properly be associated only with one or the other of them. These nuances have noteworthy implications for the understanding of the concepts of health and illness in the context of Philippine society and specific Filipino cultures.

1. SAKIT

Aside from being a rough equivalent of the word "illness", *sakit* is also a Filipino word for pain. For instance, to have a toothache is to have a *sakit* of a tooth. To have a headache is to have a *sakit* in the head. To have a stomachache is to have a *sakit* in the stomach. Hence, to have a *sakit* is to have a feeling that is localized. It is to have a pain in a particular part of one's body. In this sense, one cannot just have a *sakit*. One must have the *sakit* somewhere, in a physical space that can be identified.

Having a *sakit* in a part of one's body could mean having a feeling of pain that is caused by a medical condition in that particular part of the body. To have a *sakit* in

R.-Z. Qiu (Ed.), Bioethics: Asian Perspectives, 105-112.
© *2003 Kluwer Academic Publishers. Printed in the Netherlands.*

a tooth is to have a feeling of pain due to caries, a rotten root canal, or some similar condition in that particular tooth. To have a stomachache is to feel in the stomach the effects of, for example, something that one may have eaten. Thus, to have a *sakit* is to have a feeling of pain that is attributable to an aberration in the condition of a particular part of the body.

If this were all that *sakit* consisted in, treatment would involve nothing more than doing something about the physiological cause of the disease. However, a localized pain is not exactly all that having a *sakit* means when reference is made to a disease. When somebody complains of having a *sakit*, he does not ordinarily refer to the mere having of a specifically localized pain. In saying that one has a *sakit*, one usually refers to a broader phenomenon involving well being in more general terms. But the association of disease as *sakit* with a physical locality indicates its focus on physiological processes.

This focus on physiological processes resembles Christopher Boorse's characterization of health as "statistical normality of function, i.e., the ability to perform all typical physiological functions with at least typical efficiency" (Boorse, 1977, p. 542).

For lack of technical training and expertise, the ill person may not be able to identify a specific location of the anomaly that he or she wishes to report. But the underlying assumption is that the felt symptoms are attributable to biophysical causes.

2. *KARAMDAMAN*

The word *karamdamam* literally means "feeling". My *karamdamam* is what I feel. The root word is *dama* (literally, "to feel") and, in ordinary discourse, various affixes may be introduced to indicate a large family of meanings. The basic element in these many forms of the word is "feeling". Ordinarily then, the Filipino equivalent of "Do you feel anything?" would be taken to mean "Do you feel ill?"

This is not to say that only illness can be felt, or that illness necessarily must be felt. *Karamdamam* can refer also to the experience of pleasure, love, pity, etc. Moreover, illness may exist sometimes without being noticed or felt. Nevertheless, because of the primary association of *karamdamam* with feeling, the use of the word to refer to disease or illness puts emphasis on its experiential-phenomenal character. And, because of this experiential-phenomenal character, a wider viewpoint is engendered, which situates feelings of illness within the socio-cultural context of the valued or disvalued states.

Unlike a *sakit*, a *karamdamam* is not normally regarded as localized. If a person says that he or she has a *sakit*, it would be natural for somebody else to ask for a location. But, if a person says that he or she has a *karamdamam*, it would be more natural to ask for an explanation. In the latter case, the inquiry leads to a reason or a cause. To provide a cause, one might also point to the site of a localized pain. However, that would only be one of many possible explanations. The important thing to note is that whereas *sakit* connotes a biophysical and clinically verifiable medical condition, *karamdamam* connotes a medical phenomenon that cannot be understood fully apart from the individual's feelings and sensations, as well as his or her system of social and cultural values.

The distinction that is made here between *sakit* and *karamdamam* runs parallel to a distinction made in medical sociology or anthropology between two perspectives on disease. Based on one perspective, a disease is a biophysical condition as explained in terms of germs and dysfunction. Based on another, it encompasses diverse cultural views of the nature and causation of a sick person's experience. The latter perspective proceeds from the view that "There is . . . no essential medicine. No medicine that is independent of historical context. No timeless and place-less quiddity called medicine" (Kleinman, 1995, p. 23). If so, we must observe that "Presenting cases is not merely a way of depicting reality but a way of constructing it. It is one of a set of closely linked formative practices through which disease is organized and responded to . . .The patient is formulated as a medical project" (Good, 1994, p. 80).

In contrast with *sakit* and its emphasis on the physiological, we find in illness as *karamdamam* a reference to a wider context of valuation accommodating various notions of illness explanation and causation. Even though a physiological anomaly may be a feature of the condition, we can see other aspects integrated into the whole picture: "Medicine is deeply implicated in our contemporary image of what constitutes the suffering from which we and others hope to be delivered and our culture's vision of the means of redemption" (Good, 1994, p. 86). Given this clarification, one is in a better position to understand the nature of folk medicine in general and of some cases of faith healing in particular.

3. FAITH HEALING AND THE NATURE OF ILLNESS

One of the types of faith healing practiced in the Philippines is performed by the 'tambalans' in the Waray-speaking region of the country. Galvez-Tan notes the close integration of the healing practice with religious devotion:

> A distinct characteristic of the Waray tambalans is their loyal devotion to God. It is always their practice to give advice to their patients concerning good will to fellowmen.
>
> When asked where they get their powers, they give different responses. A common response, however, is that they get their power through some supernatural participation which is always attributable to God.
>
> [One tambalan] says that all her powers to heal come from God . . . For treatment she only uses *lana* (coconut oil) with a small piece of coconut root immersed in it. This is supposedly sacred and specially blessed . . . She does not receive payment for her services but people give their donations in cash or in kind. During the course of therapy, her patients have to offer candles before the altar in her home every Wednesday until they get cured (1977, pp. 16-17).

The various diagnostic tools employed by the tambalans indicate a methodology that may be regarded as scientifically dubious when viewed in the light of mainstream medical standards, as we can see from the following examples:

> The most common form of diagnosis is pulse-taking. Through the pulse, they can tell whether the sickness is due to supernatural spirits or due to natural causes.
> ......
> Another form is the *ripa*. There are two rituals . . . The tambalan uses the egg of a chicken of the native variety. It should be freshly laid, untouched by human hands . . . The tambalan is the only one who can hold the egg. He runs the egg throughout the different parts of the body, making signs of the cross [with it]. When this is through,

the egg is broken on a clean plate and mixed with the juice of the plant *herba buena*. The tambalan can see from the mixed egg and juice certain forms or figures whereby he distinguishes whether it was a *Kahoynon* or a *banwa-anon* that caused the ailment. He also gets to know the place where such a spirit resides. ...

......

The other form of *ripa* also involves the egg being run through all the parts of the body. The egg, however, is not broken; instead it is made to stand on a plate. Questions are asked and if the egg stands on its own on a specific question asked, then it means the answer is affirmative. The questions, only answerable by yes or no, usually ask whether such and such supernatural spirits were angered (Galvez-Tan, 1997, p. 17).

These accounts of the rituals involved in diagnosis reveal a seemingly naive and implausible aspect of faith-healing that, for many, reduces the exercise to a triviality, or worse, witchcraft. The empirical link between the diagnostic apparatus and the experience of *karamdamam* is so difficult to comprehend that one is left to wonder whether what is involved is nothing more than mere trickery and deception. For how can the configuration of the combined egg and plant juices have anything to do with the causes of a person's illness?

Nevertheless, one may approach this issue from a different angle and question the necessity of the type of empirical validation that can be demanded. In the first place, an empirical validation of the diagnostic regimen as described above may not be relevant at all. The reason is that the value of the healing rituals need not be found in their content, or their physical component. Instead, the value may be integrated in the cultural symbolism and meanings within the context of the people involved. Traditional medical practitioners are able to sift through the folk imagery and read in them a kind of configuration that renders the entire picture meaningful within their culture.

Moreover, the diversionary effect of the trivial-looking procedures could merely be overshadowing other aspects of traditional medicine that have accounted for empirically established gains. For instance, Tiston provides an explanation for the success of the tambalans: "The folk medicine user, the faith healer and the *espiritista* will patiently listen to [their] patient as the latter pours out his troubles. The tambalan is generous and lavish with his sympathy which, as it often turns out, is what the patient needs in the first place" (1977, p. 33).

Perhaps the triviality that characterizes a part of the diagnostic regimen is even necessary in the sense that it puts people's minds at ease in an environment that is familiar and not threatening. To many people in the rural areas–for that matter, even in the most highly urbanized areas in economically developed countries–modern medical apparatus constitutes a puzzling threat that easily can be mishandled. The effectiveness of such apparatus depends, to a large extent, on the faith and trust that the medical team is able to generate. And where faith and trust is a vital component of the healing process, modern medicine is given stiff competition by folk medicine, with its rich medical culture and traditions.

In a way, the methodology of modern clinical medicine shifts the context of treating diseases or illnesses away from the natural living environment to the scientific security of the clinic and laboratory. On the other hand, the methodology of folk medicine strives to retain the treatment function within the environment that a patient is most familiar with. And it is not just a physical environment that is important, but the entire socio-cultural milieu in which a patient's values and traditions have their natural home. The explanation and perception of health remedies within this natural context determines whether they can be regarded as authentic (i.e., as being grounded in their cultural identity), and hence, appreciated by the patients involved.

Moreover, the explanation as to the source of the healing power of a tambalan contributes to the authenticity that is seen to characterize his or her diagnosis and remedies. The perception of his or her power as emanating from the spirits of departed relatives confirms the bonds of kinship among the members of the community. It also exemplifies the continuing social roles that need to be played, even as they transcend the world of the living.

The attitude of people towards folk healers in general also reflects a confidence in the familiar and ordinary within their community. Referring to the practice of the arbularyo (herbal medical practitioner) in the province of Batangas, Macalintal observes:

> His diagnostic technique is developed in such a way that it is familiar to the people within the area. So through this approach, the people get the assurance that the system of folk medicine is not at variance with what they already know. They can readily accept this practice because they are made to believe that it is part of their way of life and that it is the best way of coping with the present health problem (Macalintal, 1977, 39).

Quite clearly, this emphasis on familiarity in the understanding of illness as *karamdamam* underscores the importance of authenticity in traditional medical practices.

4. CAUSES OF ILLNESS

The social and cultural dimensions of disease and illness may also be understood in relation to different theories regarding the causes of these phenomena. Tan classifies these causes in the context of Philippine society as mystical, personalist and naturalistic:

> Mystical theories attribute illness to the automatic consequence of the victim's acts and behavior. Personalist theories attribute illnesses to the active intervention of sensate agents such as supernatural entities or malevolent human beings. Finally, naturalistic theories attribute illnesses to impersonal natural forces or conditions such as cold, heat winds, or an imbalance of the body's elements (Tan, 1987, p. 82).

The various theories of illness causation reflect an integrated mass of concepts and symbols that Filipinos employ to understand their natural and social environment. These concepts and symbols go far beyond the empirical confines of the biophysical.

The attribution of illnesses to such factors or entities as ghosts, sorcerers, witches, supernatural entities, ecosystem imbalance, disruptions in social relations, environmental transgressions, etc., reflects a history of varying outlooks towards the nature of human beings, their interpersonal relationships, and their responsibilities: "Theories of illness causation embody a system of labels and attributes through which Filipinos conceptualize and recognize not only health and illness but also a variety of natural and social phenomena" (Tan, 1987, p. 87). Taken together, the various theories provide a reinforcement of the idea of illness as *karamdamam*. They also affirm the value of authenticity in healing practices. They remind us of a need to situate the search for health remedies in the context of cultures, traditions and ideologies.

These theories of illness causation explain why we encounter many cases where, even in the face of life-threatening conditions, mainstream clinical remedies easily could be ignored in favor of traditional healers' prescriptions. Patients are able to see the latter in connection with their family's and their own way of life. The patients and their families are bound to accept the traditional healer's remedies because they are explained in terms of experiences and concepts that they can recognize and associate with.

On the other hand, clinical remedies are based on mainstream medical concepts and principles that are alien to ordinary individuals. Even when they effectively deal with biophysical problems, they may not be fully understood. They may not be truly appreciated as valid remedies.

5. THE INADEQUACY OF UTILITARIAN AND PRAGMATIC CRITERIA OF VALIDITY

From a purely utilitarian or pragmatic standpoint, success in medical treatment may be gauged purely in terms of clinically measurable standards. Effectiveness in bringing about physiological normality is primary.

The importance of this criterion of validity is easy to appreciate. After all, pain does not have to be anything other than itself to be unwelcome. It even appears self-evident that pain ought to be removed. More generally, a medical aberration—perhaps by definition—needs to be corrected. Something would not be called an aberration if it were not regarded as something to be straightened out. However, an aberration is definable not only in physiological terms, but also, and more importantly, in socio-cultural terms. This is the reason why utilitarian and pragmatic considerations cannot be the only gauge of the success of medical remedies.

Aside from utility, authenticity is also an important gauge of the validity of treatment. A treatment plan needs to be in tune with the worldview that defines the concept of illness:

> Our medical beliefs, symbolisms and rituals reflect a way of looking at the world around us and help us explain or rationalize the existing order of things, including our relationships in society . . . This ideological component guides our behavior, not only in terms of the type of medical intervention that we seek but also in the ways we behave as part of society (Tan, 1987, p. 10).

In health and medical practice, it is not enough that beneficial results be derived. In the first place, beneficence cannot be understood solely in terms of the maximization of pleasure or of some other good, but also in the alignment of such goods with a people's ways of appreciating the world around it.

6. TOWARDS CULTURAL INTEGRITY AND AUTHENTICITY IN MEDICINE

The distinction between modern medicine and folk medicine in the context of clinical practice tends to isolate the latter and project it as an unscientific and outmoded attempt at providing treatment. But the dichotomy may actually be an artificial one that is fostered by a fascination with the quantifiable and the clinically verifiable. This outlook is enhanced by developments in genetic technology that are tending to situate even more aspects of medical treatment in clinics and laboratories.

Such developments have the effect of relegating traditional modes of responses to illness to the periphery even when these responses have been essential in defining a culturally authentic healing regimen all along. In part, this is due to the unconvincing nature of some exotic folk remedies. However, as pointed out above, even exotic-looking folk remedies have an important role to play in traditional healing practices.

Moreover, one should not overlook the simple caring responses that add up to ensure the authenticity of healing processes. Here, we can cite almost innumerable manifestations of concern that have their meaning in our diverse cultures: sharing, helping, reaching out, having time, talking, remaining with others, being friendly, checking on one another, various expressions of sensitivity to the needs of others, visiting, inquiring, calling, touching, giving health instructions and explanations, showing interest, fixing a meal, and many others. These are all simple and ordinary gestures that may not be perceived to have anything to do with illness as *sakit*. However, the manner in which we organize these gestures enables us to respond in ways that are culturally appropriate and authentic. These responses are in keeping with the understanding of illness as *karamdamam*.

In order to promote the authenticity of healing regimens, illness needs to be viewed as an integration of *sakit* and *karamdamam*. For this reason, the recognition of culturally sensitive medical practices through the passage in the Philippines of the "Traditional and Alternative Medicine Act of 1997" is a step in the right direction. By creating the Philippine Institute of Traditional and Alternative Health Care, the Law will help to put folk medicine in its deserved place in the mainstream of health maintenance. It needs only to be emphasized that any efforts to subsume traditional medicine under modern medicine must be resisted. The criteria of validity of modern medicine cannot be permitted to supplant the authentic perspectives of traditional medical practice. Authenticity ought to be recognized as a distinct criterion of the validity of medical treatment, both in its ethical and clinical aspects.

University of the Philippines
Dilliman, the Philippines

NOTE

[1] "*Sakit*" and "*karamdamam*" have equivalents terms in non-Tagalog Filipino languages. For instance, in Bontoc, one reports a pain by saying "*insakit*" and expresses a feeling of *karamdamam* using the term "*nasakit*" (Cf. Botengan, 1994, 38).

REFERENCES

Boorse, C. (1987). 'Health as a theoretical concept,' *Philosophy of Science*, 44, 542 - 573.
Botengan, K.C. (1994). 'Bontoc concepts on sickness and death.' In: H.W. Kiley (Ed.), *Filipino Tribal Religious Experience: Sickness, Death and After-Death* (pp. 38-41). Quezon City: Giraffe Books.
Galvez-Tan, Jaime Z. (1977). 'Religious elements in Samar-Leyte folk medicine.' In: L.N. Mercado (Ed.), *Filipino Religious Psychology* (pp. 3-21). Tacloban City: Divine Word University Publications.
Good, B.J. (1994). *Medicine, Rationality, and Experience*. Cambridge: Cambridge University Press.
Kleinman, A. (1995). *Writing at the Margin: Discourse between Anthropology and Medicine*. Berkeley: University of California Press.

Macalintal, Ma.M.B. (1977). 'The psychology of the arbularyo in Batangas.' In: L.N. Mercado (Ed.), *Filipino Religious Psychology* (pp. 36-40). Tacloban City: Divine Word University Publications.

Tan, M.L. (1987). *Usug, Pasma, Kulam: Traditional Concepts of Health and Illness in the Philippines* Quezon City: AKAP.

Tiston, R.C. (1977). 'The psychology of the tambalan in Leyte.' In: L.N. Mercado (Ed.), *Filipino Religious Psychology* (pp. 22-35). Tacloban City: Divine Word University Publications.

PART III:

LIFE, DEATH, EUTHANASIA AND END-OF-LIFE CARE

CHAPTER 11

KANG PHEE SENG

CLONING HUMANS? -- SOME MORAL CONSIDERATIONS

1. THE SHEEP THAT STUNNED THE WORLD

The successful cloning of a Scottish ewe ("Dolly") by somatic cell nuclear transfer technology announced in February 1997 has rendered human cloning viable (Wilmut *et al.*, 1997).[1] This possibility has aroused the deep and widespread concern of academics, professionals, legislators, governments and the general public. After the news of Dolly became public, U.S. President Bill Clinton immediately issued a moratorium on the use of Federal funds for programs associated with human cloning research, and further requested the National Bioethics Advisory Commission to submit a report within ninety days. On January 12, 1998, nineteen European nations signed a treaty that prohibits human cloning because it is "contrary to human dignity and thus constitutes a misuse of biology and medicine."[2]

The application of cloning to human being has generated much controversy.[3] To some, this new development merely provides another choice of reproduction. To others, it represents a fundamental change of human procreation into a technical operation that will have far-reaching repercussions on many aspects of human self-understanding and inter-personal relationships. To some, this new technological breakthrough is unstoppable. To others, this calls for a halt because otherwise "we may be sleepwalking, step by step, into a world which few of us would now choose" (Glover, 1989, p. 20). Polkinghorne is right to urge that the "technological imperative, encouraging the continuing pioneering of new techniques, must be tempered by moral imperative, requiring that such techniques (be) achieved by ethically acceptable means and employed for ethically acceptable ends" (1997, p. 42). This paper discusses some of the moral issues involved in the means as well as the ends of human cloning. It explores the relation between genetic equivalence and identity equivalence, the implication of inheriting another person's genes, the effect of asexual reproduction on family structure and relationships, the risks involved in human cloning, and the question of cloning and rights.

R.-Z. Qiu (Ed.), Bioethics: Asian Perspectives, 115-127.
© *2003 Kluwer Academic Publishers. Printed in the Netherlands.*

2. GENETIC EQUIVALENCE

It is generally assumed that the genes of the clone and those of the donor will be completely identical. In fact, although the cell nucleus of the clone comes from the donor, its cytoplasm is taken from another person. The cytoplasm contains mitochondria, which in turn have their own genes. Therefore strictly speaking, the genes of the clone and the donor, though very similar, are not identical. To achieve a hundred-percent replication of genes, the cytoplasm of the donor's mother must be used in somatic cell nuclear transfer, and even then, the mother's cytoplasm would have already undergone significant changes (Ferguson 1990, pp. 22f; Wu 1997, p. 382). The identical genes of the clone and the donor therefore refer to their identical cell nuclei. However, because biological traits are basically determined by the cell nucleus, it is generally acceptable to consider the genes of the clone and the donor as identical unless precision in scientific terminology is required. In actual fact, they are not more identical than monozygotic twins who have the same mitochondria because they are born from the splitting of a single fertilized egg.[4] Thus, if the objection to human cloning is that no two persons should share the same genes, then the presence of identical twins is a sufficient counter-argument.

Some objections to human cloning arise from equating genetic equivalence with identity equivalence (Wachbroit, 1998, pp. 66-67). A cloning clinic is wrongly perceived as a photocopier where a person goes in from the left and a hundred copies come out from the right. The Chinese may equate cloning with the Monkey God's magical self-multiplication, and an uninformed person may think that a law-breaker can get away scot-free by having an alibi through her clone. They may even perceive a clone as the reincarnation of the donor. In other words, Pol Pot of Cambodia may be resurrected and Hitler immortalized.

It is necessary to emphasize that somatic cell nuclear transfer technique involves only the cloning of genes, not of minds, thoughts or virtues. Even if they have identical genes, the clone and the donor are two different persons with different identities. Thus, when a man has an affair with the clone of his wife, he is not having sexual relations with his wife. Just as monozygotic twins have identical genes but different bodies and identities, the donor and the clone are not the same person. Cloning can neither recall the dead nor immortalize the living.

To say that the clone and the donor are the same person is to ignore the influence of "nurture" on a person. Although the clone's innate qualities may be determined by genes, the environment into which she was born will make her a person different from the donor. Indeed, not only is it impossible to reincarnate the donor (a perfect replica is impossible), a lesser goal of replicating achievement (to recreate another heavyweight boxing champion, or political leader, or great thinker, etc.) may also not materialize. In other words, cloning is only the reproduction of genes, not a reproduction of the person. Just as identical twins "are given two souls by the church and two votes by the state", so will the clone and her donor (Butler and Wadman, 1997, p. 9). All clones are to be treated as full human beings.

If there were no identical twins, we might have been led into thinking that the basic quality and uniqueness of a person lie entirely in her genes, that her genes determine her identity. This would be a reductionist view of human beings: equating human individuality solely with genetic identity.[5] Hence, to the question: "Can two persons have identical genes", the answer must be an unambiguous "yes".

While citing monozygotic twins may serve to alleviate the fear of identity confusion in human clones, we must be mindful that there are fundamental differences between the two: cloning cannot be reduced to twinning. For a start, identical twins are formed naturally in the mother's womb while clones are the products of asexual reproduction and technological intervention. The genes of twins come from their natural parents, but the genes of the clone are derived almost entirely from a donor. There is no ambiguity as to who the parents of identical twins are; but for clones, the concept of parenthood is obscure and complicated, requiring much clarification.[6] Moreover, twins are rare, triplets are scarcer, and quadruplets or more are usually the results of medication. There is, however, theoretically no limit to the number of clones. Finally, the age difference between identical twins is negligible but there could be decades and even centuries between the donor and her clone.[7] These differences have important consequences and will be explored in the rest of this paper.

3. GENES WITH A PAST

Cloning cannot be simply reduced to twinning.[8] A clone is not merely a delayed twin. In fact, the relationship between twins is not applicable to the relationship between the clone and her donor. For twins, because of the negligible difference in their age, one rarely overpowers the other. Each will have her own room for development. But can a clone maintain her autonomy for independent development without being influenced by the donor?

The birth of a new life is a chance combination of her parents' genes. The genes of every child constitute a new and unique combination. To all parents, a newborn is an unknown. She begins life with a kind of "genetic independence" from the parents. She replicates neither her father nor her mother (Meilaender. 1997, p. 42). The parents watch the new life blossom. Each day is a new page of a life symphony, filled with surprises. The parents could be athletes but their child may be quiet and prefer reading; or they could be professors in mathematics but their child just wants to draw. Every child is a newborn both in name and in reality. Parents must observe with care so that they will not demand from their child according to their own wishes but will allow her to develop according to her natural inclination. Although the genes of a child come from her parents, they are unique and belong exclusively to the child. The unpredictability of a child's future means that she is given space for personal development and freedom to exercise her choice and will (Yuan, 1997, p. 64). The future of every child is therefore open. This is true also of twins.

But the same cannot be said of the clone. The genes of a cloned child are derived completely from her donor. They are not a unique new combination of two sets of genes. Some aspects of her future, her physique, disposition and talents have already been partially revealed in the life of the donor. No doubt we must avoid absolute genetic determinism. But we cannot ignore the fact that genes do determine some innate characteristics or biological traits of a person which have a significant influence on her future growth. While a clone of Carl Lewis may not be guaranteed the same success of nine Olympic gold medals, he would at least be endowed with a physique capable of producing an excellent athlete. The quality of the donor's genes is in fact very likely the main reason for cloning. The core of the issue is not whether the clone of Marilyn Monroe is Marilyn Monroe, which, as has already been pointed out, is not the case. But to Monroe's admirers, her reappearance through a clone is already her resurrection. The purpose of cloning has already been achieved. Similarly, although a mother who has lost her child is aware that it is not possible to bring back the dead, a clone reawakens fond memories of the beloved and indeed "re-presents" the latter. In other words, the future of a cloned child is not completely open because her future is clouded by the history of her genes. The past is too much with her. Her end (*telos*) is a return to the past.

In theory, a clone and the donor are two different persons. Each has her own identity. But in reality, how much room does a cloned child have to develop her own future and identity according to her own will and inclination? How much of it is dictated by the history of the genes she inherited? Will a clone of Michael Jordan (or his "cloner") be able to resist the multi-billion dollar lure of basketball games? Will the child have the true freedom not to choose basketball as a career? There are already concrete examples. Just by looking at how the offspring of Kennedy and Nehru (Gandhi) are being drawn into politics, we can envisage how much freedom a clone would enjoy in developing her future.

A cloned child is not a true newborn because her genes are copied—not new, not unique. Before she even exists, there has already been someone who has lived with those genes. To be sure, it is not that a person's uniqueness and dignity must be built upon a unique set of genes. But a cloned child's dignity is infringed because the space for her personal development is somewhat bounded by the past events of her genes. It is almost inevitable that the new life will be constantly scrutinized in relation to that of the old.

Whereas a natural child is the fruit of the loving union of her parents, a cloned child could easily have been the designed product of some individual. In the case of the former, a new life is given to the parents in the union. It is the beginning of a unique person, totally different from the parents. In the case of the cloned child on the other hand, it was the result of a programmed project realized with the aid of technology (O'Donovan, 1984; Meilaender, 1997). The genes already existed before cloning and have been selected because they have certain value or meet a certain need.[9] The pre-existence of the genes may lead to their commodification and hence the objectification of the child.[10] It is difficult, if not almost impossible, for a cloned child to have a life that truly belongs to her. Not only will she live in the shadow of the donor, she is in fact the shadow of the donor.[11] She can't declare "I am who I am" because the donor is the original and the prototype. She is reproduced in the image of the donor. Her success and failure will be compared to that of the donor. The donor is the root while she is only a new stem. She comes *from* the donor and even *for* the donor. Her development will come under cruel pressure and her individuality will be seriously violated. Given the importance we place on the value of self-determination, we must "enable persons to live their lives the way they themselves want to live these lives" (Holm, 1998, p. 162).

Having "genes with a past" in a way also infringes on the privacy of the cloned child. It is conceivable that an individual may not even want to have knowledge of her own genetic predisposition because that can seriously affect her sense of who she is.   This has been the basis of arguments for a right not to know (Chadwick, 1997, pp. 17-20).[12] The following scenario would illustrate the embarrassment and pressure a child has to bear when his genetic secrets are publicly known.

Suppose the parents of Stephen Hawking, seeing the great potential of his mathematical talent, decided to clone Hawking Junior when Hawking was 20 years old. Unfortunately, some years later Hawking was found to have amyotrophic lateral sclerosis, the symptoms of which included physical weakness and paralysis, as well as impaired speaking. When Hawking Junior was 10 years old, he knew that in 20 years' time, he would probably suffer the same fate as Hawking. Since everyone knew that Hawking Junior was a clone, they all knew that he carried the same malignant gene.[13] Imagine the pressure Hawking Junior had to bear since his childhood as a result of his genetic information being made public.

As a cloned child, the genetic predisposition of Hawking Junior would be made public when it became evident in his donor. As such, no insurance company would provide him medical coverage. Why should his genetic privacy be infringed upon?

4. ASEXUAL REPRODUCTION

Another impact of human cloning is that cloning is asexual reproduction. Up to now, a new life always begins with the union of sperm and ovum, regardless of whether the conception is aided or unaided. But in the case of cloning, life begins with the reactivation of a differentiated somatic cell under controlled conditions to perform the function of the fertilized egg. Sexual union and relationship are bypassed.

Proponents of human cloning claim that asexual reproduction is just an alternative to sexual procreation. Just as IVF provides an artificial insemination method, cloning may offer an additional choice to those who wish to have children. For some, cloning may be the only possibility for having offspring that share their own genes. Is it therefore so critical that a child must be derived from the merging of sperm and ovum, a chance encounter of such low probability that it is seemingly accidental in any individual case?

The birth of a new life begins in the loving embrace of the parents, the union of a man and a woman. The ultimate purpose of the differences between male and female is for procreation. It is only right that the birth of a new life should be firmly grounded upon a loving, secure relationship bounded by a sacred "I-Thou" covenant between man and woman (Buber, 1970). The profound mystery of life is the sacred union of two opposite sexes. Life begins when love bears fruit. When a child is born of this loving union, she gives a new dimension and direction to this "I-Thou" relationship. Should any child be deprived of this fundamental right of being created in the love and union of her parents? Should we reduce the beginning of a child's life to a mere technological project, a mechanical reproduction? As O'Donovan would have it, "The She (or He) which will spring from the I-Thou is always present as possibility, but never as project pure and simple. And precisely for that reason she cannot be demeaned to the status of artifact, a product of the will" (O'Donovan 1984, p. 17).

A child that comes from the union of a man and a woman has her own biological parents. Cloning, on the other hand, severs the beginning of life completely from this union. In such an asexual reproduction, the question of parenthood becomes blurred and even complicated. For example, if the famous Chinese actress Liu Xiaoqing, who is divorced, decided to clone herself, who would be the father of her cloned child? Is the father of Liu, who is the father of her genes, also the father of the clone? Perhaps there is no father at all? The child is a literal single-parent child. Similarly, who is the biological mother of this cloned child? Is it the surrogate mother? Is it Liu herself, both as "twin sister" and "mother"? Or is it Liu's maternal mother who is the mother of her genes? Or has the child no mother? This problem cannot be resolved merely by legislation alone. The law can specify who the guardian and thus the legal parent is, but it cannot stop the poor child from searching for her roots, the lost parent that never was. Or perhaps the whole issue as to who should be the parents of Liu's clone is meaningless. When we allow cloning to sever life from the union of man and woman, the cloned child as a technological product is deprived of the right to her own parents (O'Connor, 1998, p. 10).

Asexual reproduction will drastically change the parenting relationship and undermine the stability of the family (Shen, 1997, p. 251). Here are some scenarios:

1. Parents by instinct are usually more committed to children who are genetically related to them.[14] To nurture the young of others is a high calling. Therefore a child who has two biological parents who equally contributed to her genetic make-up is generally better off than a cloned child who has only a single biological parent. In cloning, "the usually sad situation of the 'single-parent child' is here deliberately planned, and with a vengeance" (Kass 1997, p. 23). Even when a cloned child is in a two-parent family, because only one parent is genetically related to her, this might affect the care and commitment of the other parent towards her.

2. In a divorce, the mother usually has custody of the children. But when the parents of a cloned child divorce, does custody go to the father if the child is the father's clone? (This exposes the fundamental problem of the fact that a cloned child is not biologically derived from both parents.) Will the mother be committed enough to look after the father's clone who is not genetically related to her and who increasingly resembles the husband who is now separated from her? Will the husband pay alimony for the support of his ex-wife's clone?

3. When a clone-daughter of a mother grows, her appearance will increasingly resemble the mother whom the father was madly in love with in their younger days. If the mother unfortunately dies young, will the father, who is genetically not related to the daughter, be easily attracted to her? Will he be more tempted to incestuously violate her?

If asexual reproduction deprives a child of two parents and stands to shake the very foundation of a family, is it not too high a price to pay?

5. CLONING RISKS

As expected, the early stage of cloning research has had a very low success rate. To clone Dolly successfully, Scottish scientists used more than a thousand unfertilized eggs and went through 277 developing embryos, a third of which were deformed. Moreover, some cloned lambs were born with severe and lethal birth defects. Then there are still unresolved questions about how long Dolly will live and how healthy it will prove to be.[15] In May 1999, *Nature* reported that Dolly had telomeres which were about 20 percent shorter than those of other sheep at similar age. This indicated that her chromosomes could become unstable and more prone to damage. As a result, she could age faster and be more at risk for cancer (Shiels *et al.*, 1999). No doubt as the technique matures, we can expect a higher success rate and health index from these experiments.

The cloning of humans would be much more difficult and complex than the cloning of animals. Even if the success rate is raised significantly in future mammalian cloning, experimental failures in the process of human cloning are still unavoidable and malformed embryos can be expected. What we are dealing with here is a type of research that inevitably leads to the creation of malformed embryos. In the case of animals, if the experiments should go seriously wrong, there is always the option of halting them by destroying the embryos or slaughtering the animals. Can we do the same to human embryos? Can we permit such experimentation on human beings to be carried out?[16] Can we allow the production of "experimental human beings"?

Some may argue that nearly all medical interventions and research involve risks, and thus cloning research is no different. There is a crucial difference in cloning (Ye, 1997, pp. 34-35)! Experiments in medical science are therapeutic in nature. In many cases, patients have to choose between a low success rate when accepting therapy or deteriorating conditions and death for not accepting therapy. Cloning, however, is not a therapeutic experiment. A malformed embryo or baby from a failed cloning experiment does not exist originally. It is the cloning technology that brought her into this world. Can we then say that as long as we can clone 99 beauties like Gong Li the famous actress, it is acceptable to proceed with cloning even if one turns out to be ugly and malformed? Or 99 marshal art experts like Bruce Lee even if one turns out to be crippled?

Admittedly, we have malformed embryos or babies from natural births too. But these are natural occurrences for which no one has to be held responsible. But when scientists perform a cloning procedure knowing full well that malformed embryos and babies cannot be prevented, they should be held accountable. Some may say that since the embryo does not exist before cloning, it would be meaningless to talk about protecting its rights or being responsible for its harms. That is like arguing that adolescents need not bear the responsibility of the consequences they may cause to children who are accidentally conceived when they have sex.

It is not difficult to envisage the termination of malformed embryos to reduce the number of malformed infants in the process of cloning. The morality of creating and destroying human embryos for the purpose of scientific research is itself a highly controversial issue.[17] However, even if malformed embryos are discarded, research failures resulting in malformed infants cannot be completely eliminated. This is because chances of a mature somatic cell mutating during the entire cloning process are very real. The disorders, however, may not be fully discovered at the embryo stage, and so malformed babies could still be born. Furthermore, serious problems affecting stamina, intelligence and psychology or diseases such as unusual carcinoma and early aging may develop later in the life of the clone.

During the initial, long period of experimentation of human cloning, the entire life of almost every clone will become the object of experiment. She will be under observation by scientists for the purpose of understanding the long-term effect of cloning on her and even her offspring. In other words, the purpose of her existence will be for research (Qiu, 1997, p. 68). She will be destined to live in the fear that something might have gone wrong during the process of cloning. This is the crux of the issue: Who has the right to decide that she should live in the shadow of a possible biotechnological error? Who has the right to decide that a person's entire life should be "the object of tireless scientific and public curiosity, exposed to unending physical and psychological testing as she ages, to some degree a laboratory subject?" (Byers 1997, p. 73).

Even when the cloning technique is nearly perfected, a failure rate will still be inescapable, however small that may be. Allowing even one malformed infant to appear at any stage of cloning is morally unacceptable. Although a human clone is born from asexual reproduction, he is by no means an inferior or a second class human. A clone is a person in totality, and he is entitled to human rights. The fundamental rights of a clone must be respected and safeguarded by medical science, legislation and society. The dignity and rights of a person must not be violated; a person must not be created for the benefit of some third party, as means to an end.

6. CLONING AND RIGHTS

Proponents of human cloning often assume that the right to procreate is a basic human right. Therefore, they argue, as long as it is technologically feasible, anyone who so desires should be allowed to procreate via cloning. However, two recent cases in the West show that this right to procreate has been perhaps pushed too far. The first case involves an English woman who successfully won the legal battle for the right to use her dead husband's frozen sperm without his consent and became pregnant more than three years after his death (*The Sunday Times*, 1998). In another case in the U.S., a doctor extracted the sperm of a deceased man and used it to help his widowed wife get pregnant (*The New Scientist*, 1998). Leaving aside the moral issues, the extreme individualism and liberalism underlying the above two cases are not what an Oriental society will easily subscribe to. Does an individual's right to reproduce have priority over all other concerns? Should not this right be subservient to the interests and rights of the cloned child? Should this right be defended even if it threatens our basic family relationships and structure?

In traditional legal cases, we are concerned with the welfare and rights of existing, identifiable persons. In the new biotechnology developments, the legal issues and moral dilemmas are much more complex, for here we are dealing with the unborn (the pre-embryo, the embryo and the fetus) or even the unconceived. The relevant questions are: Does a potential person, "whose existence is not only remote in time but totally dependent on our choice," have rights? (Heyd, 1992, p. 11). Can a child born handicapped sue those who are responsible for her unfortunate condition? Does she have a right to be born healthy—physically as well as emotionally? Is there a right not to be born? A right not to be born a guinea pig of biotechnological experiments? Is the right *not* to be born miserable of equal weight and status with the right to be born happy and healthy?

It has often been assumed by some, such as Michael Bayles, that the unborn and the unconceived lack identity and hence cannot be harmed (Heyd, 1992, pp. 35ff). However, an unprecedented case in Israel a few years ago posed a serious challenge to this idea by affirming the rights of the potential person. The case involved a couple, both in their own name and in the name of their child, suing the doctor for giving negligent and erroneous genetic advice that resulted in the birth of a child afflicted with a serious genetic disease. In the ruling, the district court recognized the parents' standing but not the child's. An appeal was made to the Israeli Supreme Court which, in a majority decision, granted the child standing as well (Heyd, 1992, p. 27; Heyd, 1986).

It is important to note that the difficulty in dealing with the unconceived or "wrongful life" (suing for being born) is philosophical rather than judicial.[18] Indeed many courts have regularly brushed aside these cases "as transcending the sphere of legal discourse" (Heyd, 1992, p. 30), because as Judge Stephenson put it,

> Here the court is considering not "an ancient law" but a novel cause of action, for or against which there is no authority in any reported case in the courts of the United Kingdom or the Commonwealth.... It is tempting to say that the question... is so important that it should be argued out at a trial and on appeal up to the House of Lords. But it may become just as plain and obvious that the novel cause of action is *unarguable* or *unsustainable*... (*MaKay v. Essex Area Health Authority*, in Heyd, 1992, pp. 37-38; italics mine).

There have been in recent years an increasing number of legal cases related to claims for "wrongful life". The possibility of human cloning will no doubt add new complexities and perplexities.

Other new issues related to the right to clone will also arise. Does the clone have no right at all? Is she only to be a passive recipient of experiments performed on her? Do parents have the right to clone their child? What about the child's right? Since the genes are unique to the child, is her right being infringed upon?[19] Is cloning meant for the benefit of the parents or the child? Would the parents not have become the thieves instead of the guardians of her genes? What kind of influence will the clone have on the donor child?[20] If parents have the right to clone their child, do both parents separately own (i.e., each has the right to clone) or jointly own (where the consent of the other is required) the right to clone their child? If the parents are divorced, does the right go to the parent who has custody of the child? Is it morally right to clone a child after divorce? What happens when the child reaches adulthood? If the parents started cloning a child when he was young and kept the frozen embryo, do they lose the right to own the embryo when the child becomes an adult? Is the right to clone transferable? Is the right to clone patentable?[21]

7. TO CLONE OR NOT TO CLONE?

The application of cloning technology on humans is not merely a question of providing another choice of reproduction. The social and moral concerns arising from this possibility are so overwhelming that it has become the common concern of the international community. Perhaps the most urgent task is not to perfect the technique for human cloning so that infertile couples can have another option to procreate, much less to establish a possible scenario that would permit human cloning. As Harold Shapiro, chairman of the U.S. National Bioethics Advisory Commission and President of Princeton University frankly admitted, the Commission was not able to reach a consensus on many important moral issues after more than three months of consultation and deliberation. There are differences and divergences in opinions amongst the Commission members. Regrettably their recommendation to ban human cloning is on grounds of safety alone. Issues of life and morality have been set aside although the Commission recognizes their importance and seeks to "encourage widespread and continuing deliberation on these issues in order to further our understanding of the ethical and social implication of this technology" (National Bioethics Advisory Commission, 1997, Vol. 1, p. 110).

A pressing concern we face today is that the pace of our reflections in humanities has not kept up with fast-moving biotechnology and genetic technology (Wadman, 1997). When technology is predominantly controlled by economic and commercial considerations,[22] does our society have adequate moral wisdom, courage and strength to supervise and guide its development and application? Do we just blindly allow whatever is technologically possible to become reality? Will future generations be completely overwhelmed by technological development? Will cloning become a symbol of extreme human manipulation of nature?

To reject human cloning out of fear when the discussion of the topic by the international community has just begun is unjustified. But hastily to allow human cloning when the moral, social, legal and religious issues are still unresolved is also irresponsible. Because of the possible merchandising of genes and the enormous commercial value of cloning technology, we should be wary of the claim that cloning research is only a matter of the freedom of scientific inquiry, and that in the name of the sanctity of knowledge, it should not be held back. As Marshall Nirenberg, who won the Nobel Prize for his work in genetics, put it, "When man becomes capable of instructing his own cells, he must refrain from doing so until he has sufficient wisdom to use this knowledge for the benefit of mankind" (Quoted in Sherry, 1997, p. 121).

Human cloning may well turn out to be one of the things "that we *can* do that *ought* not to be done" (Ramsey, 1970, pp. 150ff; his italics).

Hong Kong Baptist University
Hong Kong

NOTES

[1] In the rest of this paper, mention of human cloning technology will refer specifically to the asexual reproduction of a human being by somatic cell nuclear transfer. This clarification is necessary because various cloning experiments reported in the media have significant differences both in substance and technique. In fact, after the announcement of Dolly's birth, there were frequent reports by the media of successful animal cloning in various countries. Many in fact involved the cloning of the genes of pre-embryos or embryos, or the splitting of a fertilized egg or pre-embryo (Singer and Wells, 1985, p. 137). Others involved the cloning of non-mammals. These achievements, though significant, are very different from the cloning of a grown mammal by somatic cell nuclear transfer. For a scientific account of somatic cell nuclear transfer technique, see Rossant (1997).

[2] Although the signing of the treaty was planned months earlier, it took on a greater sense of urgency when a Chicago physicist, Richard Seed, announced on January 8 that he would clone a child within two years (Associated Press, 1998b).

[3] Dissenting voices come from all sectors, not least from scientists themselves. Even Dolly's creator, Dr. Wilmut, has said that he "would find it offensive to clone a human being" (Kass, 1997, p. 19). In another interview, he stated categorically that "we (the Roslin Institute and the PPL staff) would find it ethically unacceptable to think of doing that" (cloning humans). In fact, he was of the view that "any kind of manipulation with human embryos should be prohibited" (Wilmut, 1998, pp. 50ff). For discussions of the impact of human cloning in the Chinese literature, see Lin (1997, pp. 233-267) and Yuan (1997). Wilmut's work is supported by a biotechnology company, PPL Therapeutics in Edinburgh, which plans to use the patented cloning technique to produce animals that will secrete valuable drugs in their milk (*Scientific American*, 1997).

[4] Guo (1997, p. 36) errs in claiming the complete genetic identity between a clone and her donor, and that it is more so than between monozygotic twins.

[5] For a critique of genetic reductionism, see Jochemsen (1997).

[6] Dolly has no biological father, although she has three mothers who provided respectively the cell nucleus, cytoplasm and uterus.

[7] Some proponents of cloning would like to see "an Einstein appearing every 50 years or a Chopin every century. It would be better still if we could be guaranteed not just an Einstein but the Einstein" (Kiuger, 1997).

[8] For an example of identifying cloning with twinning, see Bailey (1997).

[9] It is wrong to clone a child on eugenic grounds. However, in a capitalist society this is probably the greatest driving force behind the development of cloning technology. According to Pence (1997, pp. 101-102), it is indeed the best argument for cloning so that the child can have better genetic advantages. While Singer and Wells (1985, pp. 145-148) tend to support only "single cloning" of an individual, they too reveal their eugenic bias in advocating the "multiple cloning" of "exceptionally gifted people".

[10] It seems ironic that while Federal funding has been prohibited from human cloning research, only an appeal was made to the private sector to exercise self-restraint and to adhere to the intent of the Federal moratorium on human cloning (National Bioethics Advisory Commission 1997, volume 1, p. 109; also pp. 96-99). Given the potential commercial value in cloning research, scientists may in fact turn to private sectors for financial support. This would result in an even more undesirable scenario of cloning research being in the hands of the commercial sector.

[11] "It is important to note that the 'life in the shadow argument' does not rely on the false premise that we can make an inference from genotype to (psychological or personality) phenotype, but only on the true premise that there is a strong public tendency to make such an inference" (Holm, 1998, p. 162).

[12] "The Danish Council of Ethics has argued that ... there may come a point at which so much information is forthcoming that it may become an intrusion into the individual's private sphere, and at that point genetic screening is indefensible" (Chadwick 1997, p. 19; see also Danish Council of Ethics, 1993).

[13] While the causes of amyotrophic lateral sclerosis are yet to be determined, it has been found that approximately 20% of the familial amyotrophic lateral sclerosis have a specifically identified genetic mutation.

[14] "A television documentary chronicled an incidence where the father had wanted children 'of his own' and had assumed that it was his sperm that were being used to inseminate his wife. When twins were born with a blood type that indicated he could not be the biological father, he was devastated. Eventually, husband and wife separated because he could not reconcile himself to raising children who did not carry his genes" (Lebacqz 1997, p. 56, footnote 9).

[15] The first calf cloned in Japan died within hours after its birth (Associated Press, 1998c).

[16] Bob Edwards, whose work led to the birth of the world's first "test-tube" baby, Louise Brown, was reported to have said, "I abhor the idea of experimenting with life unless there are very good reasons for it" (The Sunday Times, 1997).

[17] The debate on the question of whether or not we should experiment on embryos started much earlier on and still continues. Dyson and Harris (1990) provide a good collection of divergent views by contributors coming from different backgrounds including law, medicine, theology and philosophy. Although the National Bioethics Advisory Commission recommended the banning of human cloning using somatic cell nuclear transfer technique as well as planting embryos into women's wombs for nurturing lives, they endorsed cloning experimentation which involves the reproduction of embryos and their termination (National Bioethics Advisory Commission, 1997, vol. 1, pp. 108-109). For those who hold respect for embryo as individual life, this is completely unacceptable. See Ramsey Colloquium (1995) and Shen (1997, p. 246) for views against embryo experimentation.

[18] David Heyd summarizes the wrongful life cases into three categories: the neonatal, the prenatal, and the preconceptive. Potential legal cases pertaining to cloning will fall into the third category (Heyd, 1992, pp. 26-35).

[19] While a child does not have the right to demand her genes to be unique (Chadwick, 1982), it does not follow that the genes which are now uniquely hers are up to her parents (or guardians) to clone as they wish.

[20] Amongst the numerous articles written on the morality of cloning, the discussion on the rights of a child donor is most neglected.

[21] Under existing U.S. patent laws, "whatever people can engineer, including living material, can be patented." (Sherry, 1997, p. 113) Different viewpoints on the patenting of biomedical products including human cells can be found in Annas (1990), Kimbrell (1994), Raines (1992), and Swain and Marusyk (1994).

REFERENCES

Sources in Chinese (in Hanyupinyin):

Guo, Zhaojiang (1997). 'Moral reflections on the cloned sheep (Dui kelong mian yang de lunli sikao),' *Chinese Medical Ethics* (Zhongguo Yixue Lunlixue), 53, 34-38.
Lin, Ping (1997). *Cloning Shocks* (Kelong Zhenhan). Beijing: Economic Journal's Press.
Qiu, Renzong (1997). 'Cloning technology and its moral implications (Kelong jishu ji qi lunli hanyi)', *Moral Studies* (Lunlixue), 8, 64-69.
Shen, Qingsong (1997). 'Moral considerations of human cloning (Lun fuzhiren de lunli wenti),' *Philosophy Magazine* (Zhexue Zazhi), 21, 233-253.
Wu, Guangdong (1997). 'Human cloning revisited (Ye tan kelong),' *Science Monthly* (Kexue Yuebao) 28, 380-382.
Ye, Qiaojian (1997). 'Is human cloning really morally unacceptable? (Kelongren zai lunli daode shang zhende bu keyi jieshou ma?),' *Ethics* (Lunlixue), 12, 32-36.
Yuan, Juzheng (1997). 'The Pandora's box of cloning technology: moral reflections after the shock of Dolly (Fuzhi keji de panduola hezi: Taoli zhenhan xia de lunli fansi),' *Contemporary* (Dangdai), 119, 58- 65.

Sources in English:

Annas, George J. (1990). 'Outrageous fortune: Selling other people's cells,' *Hastings Center Report* (November/December), 36-39.
Bailey, Ronald (1997). 'The twin paradox,' *Reason*, 29(May), 52-54.
Buber, Martin (1970). *I and Thou*, English Translation, 3rd ed. Edinburgh: T. & T. Clark.
Butler, Declan & Wadman, Meredith (1997). 'Calls for cloning ban sell science short,' *Nature*, 386, 89.
Byers, David M. (1997). 'An absence of love.' In: Ronald Cole-Turner (Ed.), *Human Cloning: Religious Responses* (pp. 66-77). Louisville: Westminster John Knox Press.
Chadwick, Ruth (1982). 'Cloning,' *Philosophy*, 57, 201-209.
Chadwick, Ruth (1997). 'The philosophy of the right to know and the right not to know.' In: Ruth Chadwick *et al.* (Eds.), *The Right to Know and the Right not to Know* (pp. 13-22). Aldershot: Avebury.
Chadwick, Ruth *et al.* (Eds.) (1997). *The Right to Know and the Right not to Know*. Aldershot: Avebury.
Cole-Turner, Ronald (Ed.) (1997). *Human Cloning: Religious Responses*. Louisville: Westminster John Knox Press.
Danish Council of Ethics (1993). *Ethics and the Mapping of the Human Genome*. Copenhagen: Danish Council of Ethics.
Dyson, Anthony & John Harris (Eds.) (1990). *Experiments on Embryos*. London: Routledge.
Evans, Donald (Ed.) (1996). *Conceiving the Embryo: Ethics, Law and Practice in Human Embryology*. The Hague, The Netherlands: Martinus Nijhoff.
Ferguson, Mark W.J. (1990). 'Contemporary and future possibilities', In: A. Dyson & J. Harris (Eds.), *Experiments on Embryos* (pp. 6-26). London: Routledge.
Glover, Jonathan (1989). *Ethics of New Reproductive Technologies: The Glover Report to the European Commission*. DeKalb, Illinois: Northern Illinois University Press.
Heyd, David (1986). 'Are "wrongful life" claims philosophically valid?' *Israel Law Review*, 21, 574-590.
Heyd, David (1992). *Genetics: Moral Issues in the Creation of People*. Berkeley: University of California Press.
Holm, Søren: 1998, 'A life in the shadow: One reason why we should not clone humans,' *Cambridge Quarterly of Healthcare Ethics*, 7, 160-162.
Jochemsen, Henk (1997). 'Reducing people to genetics.' In: John Kilner *et al.* (Eds.), *Genetic Ethics: Do the Ends Justify the Genes?* (pp. 75-83). Grand Rapids: William B. Eerdmans.
Kass, Leon R. (1997). 'The wisdom of repugnance,' *New Republic*, 216(22), 17-26.
Kilner, John F. *et al.* (eds.) (1997). *Genetic Ethics: Do the Ends Justify the Genes?* Grand Rapids: William B. Eerdmans.
Kimbrell, Andrew (1994). 'Biological patents affront human values', In O'Neill (ed.), *Biomedical Ethics: Opposing Viewpoints* (pp. 17-24). San Diego, CA: Greenhaven.
Kiuger, Jeffrey (1997). 'Will we follow the sheep?' *Time*, 149(March).
Lebacqz, Karen (1997). 'Genes, justice, and clones.' In: R. Cole-Turner (Ed.), *Human Cloning: Religious Responses* (pp. 49-57). Louisville: Westminster John Knox Press.

Meilaender, Gilbert (1997). 'Begetting and cloning (Paper Presented to the National Bioethics Advisory Commission on March 13, 1997)', *First Things*, 74(June/July), 41 - 43.

National Bioethics Advisory Commission (1997). *Cloning Human Beings,* 2 Volumes, [Volume 1: *Report and Recommendations of the National BioethicsAdvisory Commission;* Volume 2: *Commissioned Papers*]. US Government Printing Office: Rockville, Maryland.

Nelson, J. Robert (1994). *On the New Frontiers of Genetics and Religion.* Grand Rapids: William B. Eerdmans.

New Scientist (1998). 'Beyond the grave.' July 18, 159, 5.

O'Connor, John (1998). 'Human cloning would be unethical.' In: P. Winters (ed.), *Cloning* (pp. 9-12). San Diego, CA: Greenhaven.

O'Donovan, Oliver (1984). *Begotten or Made?* Oxford: Clarendon Press.

O'Neill, Terry (Ed.) (1994). *Biomedical Ethics: Opposing Viewpoints,* San Diego, CA: Greenhaven.

Pence, Gregory E. (1997). *Who's Afraid of Human Cloning?* New York: Rowman & Littlefield.

Polkinghorn, John (1997). 'Cloning and the moral imperative.' In: R. Cole-Turner (Ed.), *Human Cloning: Religious Responses* (pp. 35-42). Louisville: Westminster John Knox Press.

Raines, Lisa J. (1994). 'Biological patents promote progress.' In: T. O'Neill (Ed.), *Biomedical Ethics: Opposing Viewpoints* (pp. 17-24). San Diego, CA: Greenhaven.

Ramsey, Paul (1970). *Fabricated Man.* London: Yale University Press.

Ramsey Colloquium (1995). 'The inhuman use of human beings,' *First Things*, 49(January), 17-21.

Resnik, David B. (1998). *The Ethics of Science: An Introduction*, London: Routledge.

Rossant, Janet (1997). 'The science of animal cloning', Paper Commissioned by National Bioethics Advisory Commission. In: National Bioethics Advisory Commission Eds.), *Cloning Human Beings* (Volume 2), B1- B21.

Sherry, Stephen F. (1997). 'The incentive of patents.' In: J. Kilner *et al.* (eds.), *Genetic Ethics: Do the Ends Justify the Genes?* (pp. 113-123). Grand Rapids: William B. Eerdmans.

Shiels, *et al.* (1999). 'Analysis of telomere lengths in cloned sheep,' *Nature*, 399, 316-317.

Singer, Peter & Wells, Deane (1985). *Making Babies: The New Science and Ethics of Conception.* New York: Charles Scribner's Sons. [Revised Edition of *The Reproduction Revolution: New Ways of Making Babies,* Oxford: Oxford University Press, 1984.]

Swain, Margaret S. & Marusyk, Randy W. (1990). 'An alternative to property rights in human tissue,' *Hastings Center Report* (September/October), 12-15.

Wachbroit, Robert (1998). 'Ethical concerns about cloning are misplaced.' In: P. Winters (Ed.), *Cloning* (pp. 66-69). San Diego, CA: Greenhaven.

Wadman, Meredith: 1997, 'Bioethics: business booms for guides to biology's moral maze,' *Nature*, 389, 58.

Wilmut, Ian (1998). 'Animals cloning experiments will be beneficial to humans (interviewed by Andrew Ross).' In: P. Winters (Ed.), Cloning (pp. 49-53). San Diego, CA: Greenhaven.

Wilmut, Ian *et al.* (1997). 'Viable offspring derived from foetal and adult mammalian cells,' *Nature*, 385, 810- 813.

Winters, Paul A. (Ed.) (1998). *Cloning.* San Diego, CA: Greenhaven.

Sources on the Internet

Associated Press: 1998a, 'Japanese scientists clone two calves,' *http://www.msnbc.com/news/178094.asp*

Associated Press: 1998b (January 12), 'Some nations sign human cloning ban',
    http://search.washingtopost.com/wp-srv/WAPO/199801 12/

Associated Press: 1998c (July 25), 'Japan's first cloned calf dies',
    http://dailynews.yahoo.com.headlines/ap/intemationallstory.htlm?s=v/ap/980725

Scientific American: 1997 (March 3), 'A clone in sheep's clothing',
    http://www.sciam.com/explorations/030397c1one/O3O397beards.html

The Sunday Times: 1998 (June 28), 'Diane Blood pregnant after three-year battle',
    http://www.the-times.co.uk/news/pages/sti/98/06/28/stinwecon01001.html?1654471

CHAPTER 12

KAZUO TAKEUCHI

BRAIN DEATH CRITERIA IN JAPAN

1. INTRODUCTION

In 1901, Harvey Cushing (1902, pp. 375-400) reported the first case of brain death
in which spontaneous respiration stopped 23 hours prior to cardiac arrest. Since then,
the state of brain death has not been uncommon in neurosurgical practice, especially
with the development of modern medicine, such as intensive care and reanimation
techniques.

In 1968, soon after the first and only occasion of a heart transplant in Japan, the
Japanese EEG Society's *Ad Hoc* Committee on Brain Death was established. At that
time, the committee adopted the following declaration: "Brain death is considered to
be a state represented by irretrievable loss of cerebral function including that of the
brain stem as well as that of the cerebral hemispheres." In 1974 this committee
published criteria based on the results of retrospective clinical studies of some 200
cases of brain death collected from several neurosurgical centers, for the
determination of brain death solely in cases of acute primary gross organic damage
of the brain (Ueki, *et al.*, 1973, pp. 104-105). These criteria have since been widely
followed mainly among Japanese neurosurgeons.

Our mandatory criteria for the determination of brain death are as follows:

Deep coma
Apnea
Bilateral dilated pupils
Absent pupillary and corneal reflexes
Abrupt fall in blood pressure with persistent hypotension
Isoelectric EEG
Duration: 6 hours

R.-Z. Qiu (Ed.), Bioethics: Asian Perspectives, 129-134.
© 2003 Kluwer Academic Publishers. Printed in the Netherlands.

2. BRAIN DEATH STUDY GROUP OF
THE MINISTRY OF HEALTH AND WELFARE

Prerequisites governing Japanese criteria were not pertinent for non-neurosurgical instances of brain death. In 1983, therefore, a new brain death study group was organized under the sponsorship of the Ministry of Health and Welfare to carry out a reevaluation of the previously adopted criteria and an epidemiological study of brain death in Japan. Of course, during the 10 years following the publication of the previous criteria, there were rapid advances in the field of medicine, such as the introduction of computerized tomography and so on. In addition, in many Western countries, a second or third generation of criteria, generally official criteria approved by the government medical council or parliament, had been published.

During a 6-month period beginning March 1, 1984, 217 neurosurgical and neurological clinics and emergency services throughout Japan reported 718 brain deaths caused not only by primary lesions but also by secondary lesions, and diagnosed as such by means of the Japanese criteria excepting "abrupt fall of blood pressure followed by persistent hypotension." A further 585 brain death cases were registered during the same period, but no data were available. These figures obtained in the multi-institutional study suggest that there are at least 2,600 cases of brain death in Japan during a 12-month period. Therefore, the incidence of brain death in Japan is about 0.4% of total deaths.

The data from the 718 cases covered in this study were pooled and analyzed. A common distribution regarding age and sex revealed a peak in the 5th through 6th decades of life, with males slightly predominant. The most common primary diagnosis was cerebrovascular disease, with primary lesions accounting for 92% of cases and secondary brain lesions for only 8%. Intracranial hemorrhage was most common among cerebrovascular disease, while cerebral contusions accounted for the majority of head injury cases. Cerebrovascular disease was more frequent after the 5th decade of life, while secondary brain lesions and brain tumors were more common among the younger generations.

A description of the neurological findings at the time brain death was diagnosed, including apnea and fixed and dilated pupils, clinically present for 6 hours, also revealed variations in the size of pupils. The diameters were less than 4 mm in 20% of the cases and 5-6 mm in 56% of the cases. The level of consciousness, as classified in accordance with the Japanese coma scale, was 300 in 97.6% of all cases, and as classified by the Glasgow coma scale, 3 in 94.2% of all cases. There was an absence of cephalic reflexes in nearly 100% of the cases. However, superficial reflexes and deep tendon reflexes were preserved in some cases, and about 5% of all the cases demonstrated Babinski's reflex. Ancillary tests for a more objective determination of the state of brain death other than the EEG were rather uncommonly performed at that time. An isoelectric EEG was reported in 97.5% of the cases examined. Other ancillary tests were carried out only in selected cases, and there were no uniform findings as is the case with the EEG. The absence of a brain-stem-evoked response and an angiographic non-filling phenomenon were observed in about 80% of cases so examined. CBF studies and CT had been performed in only a few cases.

There was no improvement in any of the 718 cases considered and included in this study. In 552 cases, those in whom artificial ventilation was discontinued being excluded, the average length of the time between confirmation of the state of brain death and the moment of cardiac asystole was about 4 days. About 90% of brain dead patients died within 8 days of confirmation, and the longest period of a brain dead state was 82 days. Among those cases in which the state of brain death prevailed for more than 15 days, those resulting from secondary brain lesions showed significantly longer periods of brain death than those resulting from primary brain lesions. Autopsies were performed in about 20% of the total number of cases. Brain edema was detected in about 90% of these autopsied cases, but cerebral herniation and autolysis were observed in only 70-80% of such cases.

The conclusion of this collaborative study of brain death and re-evaluation of the Japanese criteria published in 1974 is that the former criteria are still generally reliable. However, some changes have been made, and clinical guidelines and new revised criteria for determination of brain death have been adopted as follows (Takeuchi et al., 1987, pp. 93-98):

> Prerequisite: Known irreparable organic brain lesion, detected by computerized tomography.
> Exclusion: Children under 6. Hypothermia. Drug intoxication. Endocrine and metabolic disorders.
> Criteria: Deep coma, 300 - Japanese coma scale. 3 - Glasgow coma scale.
>   Apnea, confirmed by apnea test.
>   Bilaterally fixed pupils larger than 4 mm in diameter.
>   Absent corneal, ciliospinal, oculocephalic, vestibular, pharyngeal, and cough
>   reflexes.
>   Isoelectric electroencephalogram.
> Duration of observation: 6 hours, or longer if necessary.

Although there may be no uniformity, the new criteria of brain death are rapidly coming into popular use. However, not a few brain dead patients are still being maintained on a ventilator, because relatives cannot accept the reality of death and medical personnel fear legal repercussion if they insist on discontinuing cardio-pulmonary care. Therefore, heart or liver transplantation from brain dead donors has not been performed since then.

3. BRAIN DEATH AND ORGAN TRANSPLANTATION COMMITTEE ON BIOETHICS OF THE JAPANESE MEDICAL ASSOCIATION

In 1988, the Brain Death and Organ Transplantation Committee on Bioethics of the Japanese Medical Association published their final report, which referred many discussions to the interim report of the committee from medical and non-medical professions. The summary of the final report is as follows (Takeuchi, 1990, p. 82-84):

> 1. Death of the brain, namely irreversible loss of whole brain function, is regarded as death of a human being.
> 2. Diagnosis of brain death should be made by the criteria formulated by the Ministry of   Health and Welfare's Brain Death Study Group as a minimum requirement; basic   procedures may be adopted by each committee in bioethics of the local medical institution. By this means there should be no ambiguity regarding the diagnosis, which should be prudent and certain.
> 3. It is appropriate at the present time to diagnose the death of an individual by brain death with the consent of the patient and family, respecting the will of the patient and family.

According to the final report, the principles of our new criteria are as follows:

1. The procedural aspects are clearly defined.
2. Sufficient attention is given to avoiding misdiagnosis. The contents of the new criteria are more strict than those of foreign countries.
3. The procedure is relatively simple so that it can be performed easily at the bedside.

Then, in the beginning of 1992, a final report issued by a Japanese Government Provisional Committee for the Study on Brain Death and Organ Transplantation stated that organ transplantation from brain-dead donors is acceptable in Japanese society. A minority of the council's members is still uneasy about such recognition. Despite compilations of formal criteria, there still are in Japan and many developed nations continuing discussions regarding the failure of some to understand the concept of brain death, which in turn results in disagreement even among physicians and co-medical personnel as to the criteria for brain death (Youngner, *et al.*, 1989, pp. 2205-2210; Bernat, 1992, pp. 21-26).

4. RECENT ARGUMENTS AGAINST BRAIN DEATH IN JAPAN

In Japan, the main points of existing dissent on brain death can be summarized as follows:

1. Clinical death of whole brain function or patho-morphological death of the entire brain?
2. Whole brain death or brain stem death?
3. Emphasis on vital signs and neurological findings or introduction of ancillary tests for objective evaluation of EEG, BSEPs and CBF?

Furthermore, prerequisites and exclusions (including temporal factors) for the application of such criteria often are given insufficient consideration in determining brain death.

As for the concept of brain death, we should not forget that "brain death is clinical diagnosis," as is the traditional death of human being based on the well-known triad including cardiac asystole. Morphological diagnosis of the state of brain death at the bedside is nearly impossible. None of the brain death criteria heretofore established in the world required pathological confirmation. Some opponents, including a well-known critic in Japan, tenaciously insist that brain death is not so-called "functional death of the brain" but rather should be referred to as "organic death of the brain"— in other words "necrosis of the entire brain", both of which sound strange for specialists (Tsukamoto, 1992, pp. 24-38; Ohnuki-Tierney, 1994, pp. 233-254; Feldman, 1995, pp. 265-284).

In 1988, during the convention of the Eurasian Neurosurgical Academy in Bangkok, a round-table discussion on brain death chaired by Jan Brihaye of Belgium and A. Earl Walker of the United States concluded that "brain stem death means irreversible brain damage, incompatible with life". But the following statement also was made: "the concept of brain death still remains unclearly defined. The means of determining brain stem death and whole brain death are still not generally agreed upon" (Brihaye *et al.*, 1990, pp. 75-80). Except for a few neurologically oriented physicians, namely, about 26% of the executive members of the Japanese Neurological Society, the concept of brain stem death has not been accepted yet in Japan (lgata, 1991, pp. 922-933).

In order to obtain more objective findings of brain death, various ancillary tests have been introduced. However, from the viewpoints of convenience, reliability, popularization and invasiveness, the use of conventional EEG scalp leads is the only method of choice for determination of whole brain death. BSEPs and transcranial Doppler sonography, both of which have been rapidly introduced for neuro-monitoring purposes in the neuro-intensive care unit, for many reasons are still not essential examinations for the bedside determination of brain death. We must recognize that no clinical method, including laboratory tests, can inform us of the exact state of blood flow in the entire brain up to the capillary level. Moreover, no one is aware of the patho-physiological state of intracranial microcirculation in brain death. Thus, gross circulatory arrest, which until now was frequently observed by some rough methods of CBF measurement, does not always indicate complete intracranial circulatory arrest, even though from the clinical point of view it causes so-called "total brain infarction" (Shiogai, et al., 1992, pp. 205-211).

Some, however, including physicians in Japan, propose that CBF measurements be made essential for confirmation of brain death. They also argue that some remaining hypothalamo-pituitary functions (Schrader et al., 1990, pp. 239-248), unusual spontaneous movements caused by spinal reflex, respiratory-like movements (Ropper, 1984, pp. 1089-1092) and change of pupillary size observed after the confirmation of brain death (in accordance with the criteria of the Ministry of Health and Welfare's Brain Death Study Group) were controversial phenomena in the debate regarding the determination of brain death. Furthermore, such a pseudo-theoretical point of view began to cast doubt on the concept of brain death itself.

The current situation is this: Organ transplantation from brain-dead donors is still suspended. However, partial liver transplantation from living donors and kidney transplantation from cadaveric donors are frequently performed with fairly favorable results. Cadaveric donation is legally acceptable in Japan. The majority of the Japanese medical profession, especially transplant surgeons, are eagerly awaiting the time when the use of organ donation from brain-dead donors can be approved legally. Now the Japanese Parliament has approved it.

5. CONCLUSION

According to the analysis of a Japanese bioethicist (Bai, 1990, pp. 991-992), the arguments outlined above appear to have a scientific background, but the modern brain-oriented concept of the death of a human being lays the determination of brain death on sub-clinical or minute methods which are introduced theoretically apart from the clinical status. The situation in Japan is a definite example of conflict between recent medical advances and the traditional culture of Japan (Namihira, 1990, pp. 940-941). As for brain death, we still have two ways of thought represented by the expressions: "a dead brain in a living body"; and a "corpse with a good volume pulse".

Kyorin University
Tokyo, Japan

REFERENCES

Bai, K. (1990). 'The definition of death: The Japanese attitude and experience,' *Transplantation Proceedings*, 22, 991-992.

Bernat, J.L. (1992). 'How much of the brain must die in brain death', *Journal of Clinical Ethics*, 3, 21-26.

Brihaye, J. *et al.* (1990). 'Academia Eurasians Neurochirurgica; Brain death', *Acta Neurochirurgica* (Wien)105, 78-80.

Cushing, H. (1902). 'Some experimental and clinical observations concerning states of increased intracranial tension,' *American Journal of Medical Sciences*, 124, 375-400.

Feldman, E.A. (1995). 'Brain death: The Japanese controversy.' In: C. Machado (Ed.), *Brain Death* (pp. 265- 284. Amsterdam: Elsevier Science B.V.

Igata, A. (1991). 'Report on questionnaire survey on brain death and organ transplantation' (in Japanese), *Clinical Neurology*, 31, 922-933.

Namihira, E. (1990). 'Shinto concept concerning the dead human body,' *Transplantation Proceedings*, 22, 940-941.

Ohnuki-Tierney, E. (1994). 'Brain death and organ transplantation; Cultural bases of medical technology,' *Current Anthropology*, 35, 233-254.

Ropper, A.H. (1984). 'Unusual spontaneous movements in brain dead patients', *Neurology*, 34, 1089-1092.

Schrader, H., *et al.* (1990). 'Changes of pituitary hormones in brain death,' *Acta Neurochirurgica* (Wien), 105, 239-248.

Shiogai, T. *et al.* (1992). 'Supratentorial circulatory arrest monitored by TCD in impending brain death: Diagnostic significance in loss of brain function.' In: M. Oka *et al.* (Eds.), *Recent Advances in Neurosonology* (pp. 205-211). Amsterdam: Elsevier Science B.V.

Takeuchi, K. *et al.*: 1987, 'Evolution of criteria for determination of brain death in Japan', *Acta Neurochirurgica* (Wien) 87, 93-98.

Takeuchi, K. (1990). 'Evolution of criteria for determination of brain death in Japan,' *Acta Neurochirurgica* (Wien), 105, 82-84.

Tsukamoto, Y. (1992). 'Arguments about the criteria of brain death in Japan', *Journal of Behavioral and Social Sciences*, 41, 24-38.

Ueki, K. *et al.* (1973). 'Clinical study on brain death: A collaborative study in Japan', *Excerpta Medica ICS* 239, 104-105.

Younger, S.J. *et al.* (1989). '"Brain death" and organ retrieval; A cross-sectional survey of knowledge and concepts among health professionals,' *Journal of the American Medical Association*, 261, 2205-2210.

CHAPTER 13

MING-XIAN SHEN

TO HAVE A GOOD BIRTH AS WELL AS A GOOD DEATH: THE CHINESE TRADITIONAL VIEW OF LIFE AND ITS IMPLICATIONS FOR MODERNITY

"Anyone who has a good birth as well as a good death fulfills the humanity." This is the ideal formulated by Xun Zi (313-238 B.C.), a well-known Chinese philosopher in the period of Warring States. He said; "Birth is the beginning of human life, while death is its end. If both take place perfectly then the one concerned has truly fulfilled the humanity and this is the ideal journey of life" (*Xun Zi*). This viewpoint has deep significance. It is relatively easy to suggest a good birth and a meaningful life, and it can be understood and accepted by people without much difficulty, but to suggest a good, meaningful death is something quite different. From the above perspective supported by Xun Zi, death and the way of death is a way of being human and the way to ideal life. Thus it not only admitted that death was the natural and inevitable end of human life but also affirmed the value of death.

Of course, the idea of Xun Zi is not an isolated one. Precisely speaking, the view is the reflection and representation of ancient Chinese wisdom on life and death. Just as ancient thinkers of other nationalities did in the past, ancient Chinese thinkers were much concerned with living conditions and the value of life and produced a series of thoughts full of wisdom on human life and death, some of which still remain brilliant and are worthy of being properly studied and displayed. Unfortunately these precious ideological and cultural resources have not yet gained appropriate attention and been properly interpreted in the contemporary context. Today as we stand at the beginning of the 21st century and mankind is faced with many tough challenges, the time-honored Oriental wisdom is more and more cherished by people throughout the world.

No doubt, bioethics in Asia should represent the tradition and characteristics of Asia. For this purpose I lay emphasis on the research of the view of life in ancient China and make creative alterations to it so that it will become an important ideological resource and an integral part of Asian bioethics and reveal its implications for modern time.

R.-Z. Qiu (Ed.), Bioethics: Asian Perspectives, 135-147.
© *2003 Kluwer Academic Publishers. Printed in the Netherlands.*

1. BASIC VIEWPOINTS

In order to make reasonable and full use of this precious historical treasure, it is essential to understand the basic viewpoints on life and death in ancient China. Here I list them as follows:

1.1. The Respect for Life and the Supremacy of Human Life

According to the record of the ancient medical book *Ten Questions* (excavated from the Ma Wang Dui tomb of the Han Dynasty in Hunan), there was a dialogue between the two most prestigious sage kings, Yao and Shun, in ancient China. Yao asked: "What is most valuable?" Shun replied: "Life." Shun definitely affirmed that life was the most precious and most valuable among all other things under heaven. In my opinion, the dialogue between Yao and Shun is not only the admonition for doctors to love and protect life but also the teaching for all their subjects. Furthermore, the idea of life's supremacy does not belong to a particular school, but was upheld by the whole Chinese people as their common belief. Why should we say so? Because these two kings were the revered leaders of Chinese people and also their spokesmen. What they said was regarded as the law and "the King's directive", which did not allow any doubt. Later in *Yi Zhuan*, this belief was further summarized in theoretical language: "Life is the grand virtue of heaven and earth" (*Yi Zhuan: Xi Ci Xia*). According to the understanding of the ancient Chinese, everything in the universe was alive, with constant vitality. The heaven is the positive, the male, and the earth is the negative, the female. The heaven and the earth interacted with each other, and from this interaction everything was given birth and developed. In this way the great world full of varied colors and vitality was formed. What was described in *Yi Zhuan* really is the picture of creation. If there had been no life, the world would have been bleak and monotonous, and what kind of meaning would have been left to it? Therefore, life is the inherent nature of heaven and earth and also their grand virtue. We should show our sincere gratitude to heaven and earth for endowing us with life, and follow their virtues to respect and love life. Only in this way can one become a man of virtue.

Life is the most precious thing under heaven, especially the cherishing of human life. For the thinkers of ancient China, the human being is "the spirit of all things", and among all living things, it remains on the highest level and is the most valuable. As Confucius said: "Between the heaven and the earth man is the most treasured being" (*Analects of Confucius*). We surely ought to cherish life, especially human life. In this respect, the well-known doctor of the Tang Dynasty, Sun Simiao, gave us the most definitive statement. He wrote in the preface of his book *Qian Jin Yao Fang* (*Golden Prescriptions*) that the most important thing was human life, which was thousands of times more valuable than gold. It was the benevolent wish to save human life that prompted him to be engaged in medicine and to spread medicine. His book, *Golden Prescriptions*, was deeply meaningful and had far-reaching influence.

In the light of the subject we are concerned with, the reason that Sun Simiao is worth attention is that his thought gathered the elements of Confucianism, Taoism and Buddhism. In his *Golden Prescriptions* the evidence of Confucianism, Taoism and Buddhism is very clear. The way of benevolence and justice undoubtedly pointed to Confucianism, since benevolence and justice are its basic concepts. When his disciples asked Confucius what benevolence was, he summarized his reply as "love human beings". Featuring a concern for human beings, Confucianism maintains the view that one should treat others in the way he treats himself, and not impose on others what he himself does not want. If one wants to gain a position in society, he should also let other people do so: If one would like to grow prosperous, he also should let other people become so. Furthermore, one should respect the elder and love the younger, and regard other people's elders and children as his own. This is a good way to coordinate the relationship between men and to make people love each other and work for mutual benefit. In the age of Sun Simiao, Confucianism had held the dominant position for a long time. The reason he gave up his official post and practiced medicine was to practice the Confucian spirit of benevolence. He said:

> When a doctor treats his patients, he should focus all his attention, eliminate all other desires and give full compassion to relieve the suffering of people. When a patient comes to ask for help, he should treat them equally ("*pu tong yi deng*"): the rich and the poor, adults and children, relatives and friends, foreigners and natives; all should be looked upon as the same and be treated as one's kin (*Qian Jin Yao Fang: Da Yi Jing Cheng*, [*The Sincerity of Great Physicians*]).

This time-honored precept of medical morals sufficiently shows the Confucian way of benevolence and justice. Laozi of Taoism, the thinker of the later period of Spring and Autumn (6th century B.C.), said: "Human being follows earth, earth follows heaven, heaven follows the *tao*, and the tao follows Nature." (*Lao Zi*, Chapter 25). Nature here indicates not only the elements of Nature as ordinarily understood, but also the status of nature formed beyond human power. The Taoist school assumed that Nature was the supreme state attained, and it cannot and should not be changed. *Tao* is following Nature and imitating Nature. Therefore, only by following Nature can we maintain our life and also elongate our life. On the basis of such an understanding, the Taoist school developed the systematic theory and method of keeping in good health. In this connection Sun Simiao also made a great contribution. He was honored as "the true man ("*zhen ren*") of Shun", which shows his close relation to the Taoist school. "The true man" is a special concept of Taoism and occupies the high position in Taoist School. Zhuang Zi (369-286 B.C.) said: "The true man of ancient times knew nothing of loving life, and knew nothing of hating death" (*Zhuang Zi: Da Zhong Shi*). This status can not be reached by ordinary people.

As for Buddhism, though there is no clear description, it cannot be denied that benevolence is its purpose. Sun Simiao also said: "Though there is the distinction between animal and human being, they are the same with regard to life. So in my prescription, I do not use any living thing as medicine." Obviously, it is also the Buddhist thought to love life and not to use life as medicine. The age when Sun was living was the prosperous time during which Buddhism was brought into China, with wide influence. "To save one life is better than to build seven stories of a Buddha pagoda": this is the Buddhist precept that is most widespread, and it is accepted by most Chinese. It is quite natural for Sun Simiao, as one of the greatest doctors in Chinese history, to accept the Buddhist thought of respect for life, and to relieve the suffering of ordinary people with a benevolent heart.

1.2. Loving Life and Hating Death Is a Human Being's Common Sense

Why should we respect life, particularly human life, so much? The sages of ancient China thought that it was man's inborn nature or instinct to cherish life and be delighted in survival in the world. This inborn nature or instinct cannot and should not be violated. Mo Zi, the founding father of the Mo school which advocated the love of all people under heaven, said: "It is heaven's desire to live and not to like death." Most ancient sages believed in "the unity of heaven and the human being" (*Mo Zi*). The universe not only has life but also is like a big human being and the human being is like a small universe. Therefore human beings should follow the example of the universe. If heaven wants to live, so does the human being. If heaven dislikes death, so does the human being. It is both the natural gift and the human instinct. Mencius (372-289 B.C.), the second great Confucian and Xun Zi, the third great Confucian, both confirmed that it is the common nature of the human being to desire living and dislike dying. Mencius said: "Life is what I desire and death, what I dislike" (*Meng Zi: Gao Zi Shang*). Xun Zi further pointed out that "no human being's desire is greater than living and no human's abhorrence is greater than dying" (*Xun Zi: Zheng Ming*). Living is the greatest desire pursued by humans and dying is humans' extreme fear. The deeper and richer account of this point can be found in *Huang Di Nei Jing*: "It is human sentiment to detest death and be pleased in life." What is remarkable here is the sentiment and the pleasure, which involve psychological elements. Clearly this is a step further than only mentioning "desire", the physiological elements.

The above description tells us that the ancient sages developed their argument that it is a human's common sense to be pleased in living and detest dying from three aspects, physiological, psychological and natural. Among these, to draw support from heaven's authority is most Chinese. In ancient China, there was a wonderful expression, "heaven's years", which means the years endowed by heaven. So human life span is decided by heaven or is predestined. In the long journey from birth to death, no matter how much hardship and suffering one may experience, one should know his own destiny and fulfill his "heaven's years", and not regret or violate them. This is one of the deep causes of "detesting dying and being delighted in living".

From this viewpoint, is "detesting dying" and "being delighted in living" completely passive? The answer is no. The above-mentioned way of keeping healthy developed by the Taoist school is an active approach to being pleased with living. In a paper named after the fundamentals of keeping in health, Zhuang Zi said that to breathe deeply along the spine, as a baby inspires and exhales slowly, and to focus all one's attention, helps to keep one fit and help one live longer. It is essential that the way of keeping healthy should follow "heaven's law". In my opinion, the combination of not violating "heaven's years" and keeping healthy, as well as living longer, fully reveals the marvelous thought of ancient Chinese sages.

1.3. The Value of Death May Occasionally Surpass That of Life

Though detesting dying and being pleased with living is common sense, the ancient sages did not evade death or deny its value. On the contrary, they not only affirmed the value of death but also claimed that under certain conditions the value of death can surpass that of life. In this respect there are many precious ideological treasures that have not yet gained due attention. There is even the wrong impression that the ancient Chinese sages prohibited the discussion on death and denied its value, and that the Chinese unconditionally believed in the philosophy of preferring a bad life to a good death. Now it is time to erase this mistake and recover its original status. "If you have not yet known life, how can you know death?" (*Analects of Confucius*: *Xian Jin*). This word by Confucius seems to prove that Confucius evaded talking about death, but actually it's a misunderstanding of Confucius' words which he uttered in reply to the question of his disciple Zi Lu about death. If we examine the discussion between him and another disciple, Zi Gong, the problem will become clear. According to the record, one day Zi Gong said to Confucius, "I feel tired of studying and would like to take a rest by serving the king." Confucius immediately said seriously, "It is very hard to serve the king, how can you take a rest?" Zi Gong said again, "then I would never have time for rest?" Confucius replied, "After you die and tomb has been built, then you know where to rest." Hearing this Zi Gong realized that when one was living one should constantly strive and not live a meaningless life, and when dying one should die a meaningful death. By linking death to life, Confucius maintained that only by understanding life and living a meaningful life can one understand death and die a meaningful death. This is his deep insight. How can we say he avoids the discussion about death?

Confucius once pointed out clearly that benevolence was more important than life; we may not violate benevolence for the purpose of survival. On the contrary, for the benevolent cause, one should prefer giving up his own life. This was really the benevolent and noble man. Giving up life to be benevolent is always the attitude of the Confucian school towards death, which has cultivated many noble men. Mencius said that life was what he wanted, and benevolence too; but when the two were in conflict and could not be pursued at the same time, he prefered benevolence to life. Life was certainly what he desired, but he also recognized that there were aims more valuable than life. So one should not survive without significance. Xun Zi also said that since it was common sense to detest death and be pleased with life, why were there people who preferred death to life? It was not that those people had an inborn nature which favored death, but rather that under certain specific conditions, life loses its meaning (*Xun Zi*: *Zheng Ming*). So we can see from the statements of Confucius, Mencius and Xun Zi , the masters of the Confucian school, that they all definitely affirmed there are pursuits more valuable than life, and for these values and aims it is worth giving up life. In such circumstances, the value of death is undoubtedly higher than that of life.

Zhuang Zi's observations and discussions on death are probably more deep and fundamental than that of the above scholars. Lie Zi said that we cannot only regard life as a pleasure but should also recognize the suffering in life, and we cannot only detest death but should also perceive the peace in death (*Lie Zi*). So this view is deeper and more complete than that of "detesting death and being pleased with living." Zhuang Zi admired Lie Zi very much and mentioned him many times in his works. He was deeply influenced by Lie Zi's two-sided view of life and death, and put the emphasis on the development of it in another direction. Zhuang Zi considered human existence to be hardship and suffering, and death as peaceful rest and freedom, and he claimed that this was the arrangement of Nature. Furthermore, Zhuang Zi asserted surprisingly: death is not only rest but also pleasure. In his works "zhi le" (ultimate delight) in *Zhuang Zi*, he told the story about his wife's death. After Zhuang's wife died, Hui Zi went to pay his condolences, but to his astonishment, he saw Zhuang sitting on the ground, striking a porcelain basin and singing a song, as if he was celebrating a festival. This scene made Hui Zi dumbfounded: was this appropriate behavior? The blame gave Zhuang a good chance to express his view on death: 'when my wife died, I was extremely sorrowful at first, but later I reflected on it and then realized that originally there was no air and of course no life in the universe; later came air and life accompanied by death. Life and death are like four seasons running around and are natural phenomena. When man dies, he still exists in the universe and does not disappear. If I cried loudly after my wife's death I went against the way of universe. On this consideration, I stopped crying and could not help striking the basin to praise the way of universe.'

Zhuang expressed the following view with clearer language through the mouth of a skeleton. After death, one has no king to be governed by, no humble people to be troubled from, and also no business to deal with. The delight is beyond description. Zhuang said to a skeleton: 'imagine you could recover the figure of man and once again have muscle and skin, would you like that?' The skeleton shook his head with a frown: 'How can I give up the pleasure of a king and return to those busy days?' I believe most people would wonder at such marvelous statements and the fact that two thousand years ago the Chinese thinker Zhuang Zi should have so bold a view on death. It is against the background of the way of universe that Zhuang understood life and death and regarded it as an irresistible natural phenomenon. To start from the way of universe, one would not be too sorrowful regarding death and would probably admire the greatness of the way of the universe.

If we say that it is on the level of morality that Confucius and Mencius invested value in death, then Zhuang Zi stood on the level of Nature to endow life with value. In this way it is possible for ordinary people to have the value of their deaths surpass the value of their lives. This reminds us of Socrates of ancient Greece. This great philosopher faced death calmly and optimistically. He believed that after death man could free himself of worldly trouble, and so death was probably better than life. How alike this view is to that of Zhuang! I think that in confirming the value of death, Zhuang is bolder than Socrates.

1.4. To Have a Good Birth as well as a Good Death, and to Live Following Nature and End in Peace

This is the basic attitude towards death advocated by ancient sages. Xun Zi's statement (as the title of this thesis) is quite representative:

> Ritual is what is used to treat life and death carefully. Birth is the beginning of life while death, its end. If one has good birth as well as good death, the *tao* of humanity would be fulfilled. Therefore gentlemen respect birth and take great care of death. The unity of life and death is the *tao* of gentlemen and the contents of *li* (norms, rituals) and *yi* (justice) (*Xun Zi*).

The characteristics of Confucianism are displayed in the connection of *li* to life and death. As an important category of the Confucian school, *li* is the norm of dealing with various kinds of interpersonal relationship. The Confucian school claims that only by following a certain system of *li* can the peace between persons be maintained and the good social order be kept. The way of behaving as a human suggested by Xun Zi also indicates following and observing *li*. He believed that a human being, from birth to death, should always observe the system of *li* and display *li* and *yi*, reaching the goal of a good start as well as a good end. This is the way of behaving and the ideal life we strive for. Here Xun Zi's contribution lies not only in the fact that he extended the category of virtue to the course of death, thus connecting and also unifying life and death, but also in that he put forward the aim of a "good birth and good death." He also required people to take Confucian morality as their direction from birth to death and endeavor to realize ideal life, particularly the value of death.

Here I cannot help mentioning Zhuang Zi once again, because it is he who suggests the principle of "taking both birth and death easily" which is complementary to "good birth and good death". In Zhuang Zi's opinion, life and death actually are in unity, the different expression of the same course. Life is the accumulation of *qi* (vital energy), and death, its dispersion. Therefore, one ought not to be too delighted in life or too sorrowful in death. When living, keep a peaceful mind; when dying, follow the order of nature. This is the sentiment presented by heaven. Here Zhuang Zi once again returned to the natural status. He thought that looking at the complete course of life and death in the light of morality, as Xun Zi did, loses the natural way.

Zhang Zai (1020--1077) the philosopher of the Song Dynasty, synthesized both Xun Zi's and Zhuang Zi's thoughts and proposed the well-known principle of "live following nature and end in peace". Zhang also viewed life and death as nothing other than the accumulation and dispersion of *qi*. In the end, living gains nothing and death lose nothing, but he did not thereby want to treat life and death in a completely Daoist natural way. On the contrary, he thought that it is more meaningful to live a moral life as taught by the Confucian school. He advocated that a man should deal with all kinds of business according to the moral norms of the Confucian school. If one realizes that death is the dispersion of *qi*, an irresistible phenomenon of nature, and also observes the moral norms when living and is worthy of the expectation of ancestors, children and himself, one can face and accept death calmly, peacefully, and without panic when it comes. It is easy to see that the principle of living following nature and ending in peace reduce the strong moral sense of "good birth and good death" and also the strong Daoist sense of "taking both life and death easy".

The above four points are interrelated, which has constituted the basic view and attitude of Chinese people towards life and death, and has up to this point had deep influences on the Chinese view on life and death.

2. DISCUSSION

After outlining the view of life and death in ancient China, very likely we would ask how it was formed, what merits and limits it has and whether and how these ideological and cultural resources can be utilized, so that the view acquires modern significance.

The respect for and treasuring of life is the common tradition of all mankind, regardless of nation, age, or geographical location, and has a deep-rooted physiological and psychological foundation. As is said in Chinese folklore: "Even ants have a strong desire to live, why does man not value his life?" Nevertheless I will not discuss this common problem, but rather will discuss the role played by the specific ideological and cultural background of ancient China in the formation of the predominating view of life.

In my view, there are two points that have played a key role. One is the idea of fatalism and the other is *xiao* (filial piety). Despite the fact that Confucianism and Taoism belong to different schools, they both believe in fatalism. It is written in *Analects of Confucius: Yan Yuan*: "To live or die is decided by fate, and to be rich or poor, by heaven." Zhuang Zi said: "life and death, that is fate" (*Zhuang Zi: Da Zong Shi*). Both of them believed that life and death were predestined by fate, which was "heaven's fate", the revelation of heaven's intention, and therefore was holy and authoritative, and could not be changed by humans. Confucius said: "It is no use praying if you have offended heaven" (*Analects of Confucius: Ba Yu*). Zhuang Zi also said that humans cannot triumph over nature (*Zhuan Zi: Da Zong Shi*). As mentioned above, according to the almost unanimous view of ancient China, the universe is a living being just as humans are, and the human being is the product of the interaction between heaven and earth. Even the structure of the human body is the symbol of heaven. For instance the heaven is round, and so is human head. The heaven has four seasons and so man has four limbs. Since man conforms to heaven, he should follow the example of heaven. As the birth of man is the virtue of heaven and earth, human life is given the holiness of heaven and is subject to the arrangement of heaven. Mencius and Zhuang Zi reached the same endpoint by different paths. Both of them explained and defined life and death with heaven's fate. This is worth deep reflection.

Both the Confucian school and the Taoist school, while emphasizing the mysteriousness and supremacy of heaven, also recognized that heaven had another aspect which was close to man and could be utilized by him. Today, maybe, someone would think that this was a contradiction. But in ancient China, it seems rational to recognize these two aspects at the same time, which show the thinking of ancient China. Understanding this feature will provide us with an important gate that leads to the knowledge of the thoughts of ancient China.

But how did the Confucian and Taoist schools understand the other aspect of heaven? Confucius said: "Heaven does not speak, but the four seasons run around as usual and the creatures grow as usual" (*Analects of Confucius: Yang Huo*). Xun Zi said: "The motion of heaven has its own regularities. Its existence does not depend on so sagacious a king as Yao. Its destruction has nothing to do with so atrocious a king as Jie" (*Xun Zi: On Heaven*). Interestingly, Zhuang Zi made an analogy between the shift of the four seasons and human birth and death in his work *Ultimate Pleasure (Zhuang Zi)*, assuming that life and death are just the accumulation and dispersion of *qi*, that is, the change of Nature. Later it was further pointed out in *Yi Zhuan* that everything in the universe is constantly changing. Therefore if there is birth, there must be death; if there is beginning, there must be end. If you start from the birth (beginning), the original point, and look back upon death (end), you will understand that death is necessary and inevitable. This viewpoint affected Zhang Zai enormously and was used as important evidence for his principle of living following nature in peace. The universe is constantly changing; there is birth as well as death. Is it not more reasonable to face death with an easy and peaceful mind than with panic and horror? Generally speaking, the idea of heaven's fate can be used to explain most of the life-view of ancient China, including those that are superficially inconsistent.

The ritual system of filial piety is mainly associated with the Confucian school, based on Confucian authority. It is written in *Xiao Jin* (*The Book of Filial Piety*) that filial piety, as the fundamental virtue and the beginning of enlightenment, is the law of heaven and Earth, and therefore cannot be doubted. But how should one practice filial piety? In the first chapter it is written that filial piety lies first of all in being obedient and respectful to parents and elder generations. The body, the skin and the hair which come from one's parents must be taken great care of and not harmed. This is the starting point of filial piety as well as its basic requirement. In the light of this view, no doubt parents' body, skin and hair should be given more protection. Zeng Zi, one of Confucius' favorite disciples, said: "So far as parents are concerned, when they are alive, we should attend them according to *li*. When they die, we should hold a funeral and offer sacrifice to them according to *li*" (*Analects of Confucius: Teng Wen Gong Shang*). It is an essential feature of filial obedience to let parents fulfill their "heaven's years" and die with an intact body. That is to say, when their lives are at the end, we should do all our best to save them without considering the cost so that they live to their predestined age. When they die, their bodies should be kept intact. There is a story called "bending over the coffin to extinguish a fire" in a book titled *The Pictures of Twenty Four Filial Sons*. When Zhu Guogrong's mother died and the funeral was going on, the neighbor's house caught fire. Zhu had no way to put it out. What could he do? He could only bend over his mother's coffin crying to express his intention to protect his mother's dead body with his living body. It was said that this filial action moved heaven and heavy rain fell immediately. The fire went out and Zhu became the real filial son in history. Mencius said that the greatest offense to filial piety is having no offspring to extend the life of the family. Offense to filial piety is so great a sin that it can be punished severely. For practicing filial piety and fulfilling filial ritual, respecting and cherishing life is something that cannot be spared. From this perspective, preferring "a bad life rather than a good death" becomes understandable.

Next I would like to discuss the merits of the life-view of ancient China as well as its limitations. For two or three thousand years, this view of life has always affected and dominated the attitude of Chinese people towards life and death. It has had a positive effect on reproduction, and is one of the important sources of medical morality. As a doctor in a former generation said, doctors have an extremely great responsibility because patients have entrusted their lives to him. So doctors should have a benevolent and sympathetic heart, see patients as their family members, try their best to improve their medical techniques, and no matter how serious the disease is, try to save patients' lives. Obviously this requirement of medical morality is based on the view of respecting life. At the same time as it upheld the view of sacrificing life to fulfill *yi* (justice) and *ren* (benevolence), it provided an important spiritual perspective of the value of death for Chinese people.

Theoretically speaking, the life-view of ancient China has a complete system. It cherishes the value of life and also admits the value of death. It deals with the problem of death under general condition and also under specific conditions. It has physiological and psychological analysis and also accounts concerning morality and culture. It absorbed the view of the Confucian school as well as the Taoist and Buddhist schools. What is worth noticing is that the ideal of a good birth and a good death put forward by Xun Zi and the objective attitude towards death have some originality in the cultural history of the world.

Of course it cannot be denied that the life-view of ancient China has its shortcomings. For instance, the life-view of the Confucian school has overly strong moral colors, while that of Taoist school cannot get rid of relativist shadows. The Confucian school laid emphasis on the significance of life with little attention to the value of death, whereas the Taoist school did the opposite. Indeed, it is a merit of the Confucian school that it looks at life and death from a moral point of view, because life and death are indeed closely related to morality. However, it may be too much to claim that the whole course of life and death should be measured and defined by *ren* (benevolence), *yi* (justice), *li* (ritual) and *xiao* (filial piety) and that only when life is in an irreconcilable conflict with *ren* and *yi* is it possible to choose death. In the same way, it is the wisdom of the Taoist school to see the unity of life and death. But it may also be too much to obey the fate decided by heaven without attempting to create life's splendor, or to be tired of living when seeing that death will inevitably follow life.

Since the Han Dynasty, the view of life of the Confucian school has occupied a dominant position. In practice the outcome it has produced is the neglect of life quality. The Confucian school admitted the value of death only under the specific condition of "achieving benevolence" and "fulfilling justice". Otherwise life has supreme value and death has no value. Thus it necessarily led to the neglect of life quality, especially at the end of life.

Finally, it is imperative to make a creative shift of the life-view of ancient China so that it can play a more positive role in modern times and become an ideological and cultural resource of bioethics in Asia. For this shift to take place, it is necessary and important to review the merits and limits of the life-view of ancient China in theory and practice. The preparation is certainly indispensable, but it is also insufficient. Modern content should also be incorporated into it. China is a civilized country with a long history, and it is rich in cultural resources. After several thousand years of development, the Chinese people have their specific actions, thoughts, emotions, attitudes, etc., which have been effective up to now. Nihilism and radicalism in relation to culture have proved untenable in reality and history. Tradition is an enormous force that can not be cut off or wiped out. In this sense we must affirm tradition and make use of it. At the same time, culture is always in the process of development, even in cultures like the Chinese, which is very stable. The power of assimilation must be made to fit this changing social life. It is the only way to achieve cultural vitality. History also shows that conservatism in culture is untenable. In this sense we must renovate and develop tradition. For the above reason, the creative shift of traditional culture is possibly the most appropriate and feasible approach. The so-called "creative shift" is actually to absorb and make use of the reasonable contents and elements of culture and at the same time to incorporate new contents and give new interpretations so that the view will adapt itself to modernity.

As all know, a deep change has taken place in modern medicine. The progress of life science and technology make organ transplantation and assisted reproduction available. It is also possible to sustain the respiration and heartbeat of a brain-dead patient who loses consciousness, and save the life of severely defective newborns and newborns of very low birth weight (less than 500 grams). With the number of older people increasing (e.g., according to recent statistics, there are over 2 million people over age 60, which accounts for 17.7% of the population in Shanghai, the largest city in China), and the number of patients with cancer or other serious chronic diseases increasing, the problems of the care of dying people and euthanasia become prominent. Bioethics is booming with problems closely related to life and death, making people reflect on life and death.

What is life? What is death? Since human beings have the right to life, do they have the right to death? These problems, which in the past seemed to be beyond question, have now become a hot topic of discussion. Some Chinese bioethicists emphasize life quality, and attempt to bring the sanctity and quality of life into unity, while at same time trying to find a new definition of death so as to provide moral justification for the right to death of a specific group of people. Thus bioethical issues have often caused conflict with the life-view of ancient China. On one hand, the two are not in complete conflict. There are some connections between the two. For example, both respect life and admit the value of death under certain conditions. However, there are some inevitable contradictions between the two. Bioethical issues such as organ transplantation, assisted reproduction, euthanasia and so on are entirely new from the perspective of the life-view of ancient China, and may be difficult to understand and accept. Take euthanasia as example. In the view of ancient China, a human being's body, hair and skin are given by parents and must not be hurt. How then can we administer euthanasia to dying patients? Life and death are decided by fate and cannot be changed by man. Euthanasia is opposite to life, so it violates heaven's order.

In further analysis, the relationship of the life-views of ancient China and modern bioethicists is quite complicated, and it is difficult to describe it simply by arguing they are consistent or conflict. For instance, a good birth and a good death can be regarded as the goals of bioethics, but what *is* a good life and a good death? There must be a difference between modern bioethicists and ancient Chinese sages in understanding these questions. Modern bioethicists may highly praise the courage and wisdom of Zhuang Zi in the discussion of death, but will find it difficult to agree with his view of fatalism. This situation shows that towards the life-view of ancient China, we should not restrict ourselves to keeping its merits and avoiding its limitations. But should we make this view adapt to the development of modern society and modern medicine?

From the viewpoint of bioethics, I think what is most important is to introduce the concepts of a right to health, and the value of health. Although the ancient sages affirmed the significance of death, they thought about it only for the cause of *ren* (benevolence) and *yi* (justice), outside of specific medical considerations. Within the scope of medicine and the daily life, traditional medicine still prohibited any action that would harm life, to say nothing of causing death. That is to say, in the scope of traditional medicine, only the right to life and the value of life was admitted or paid attention to, and the right to death and the value of death was denied. Furthermore, the value of life was always positive and active; the value of death, if any, was always negative and passive.

Health has already been regarded as one of the basic human rights and has been written into the charter of WHO, which also set up a program aiming at the goal that "in 2000, everybody in the world will enjoy the right to health care". Apart from this, health is also one of the most important criteria for measuring social development and progress. Physical, mental and social health is gaining more and more attention from people and governments throughout the world. The introduction of the concept of the value of health and the measure of the value of life (with the value of health as the medium connecting the value of life and death) will bring great change to many life views. Life and health are in unity. Health is based on life. Without life, how can the value of life be displayed and realized? There are healthy as well as unhealthy lives. It is owing to the change of health that life gradually makes its way to death. Therefore, under specific conditions, life and health may conflict. When life's value becomes negative, death's value will become positive. In this way, the value of health not only connects the values of life and death, but also confirms death's value in the medical field. In this way, a good life and a good death is no longer the moral ideal of a few noble men, but is brought within the scope of medicine, becoming a goal that can be pursued by ordinary people. Death can be approached with the wisdom of seeking emancipation and peace and no longer as the excuse for someone being tired of life.

It is well said by the Chinese scholar of the Ming Dynasty, Wang Mingyang (1472-1528): "Man's concern for life and death comes from his inborn nature and so cannot be removed. If someone can make a breakthrough here, then he is the man who has reached the fundamentals of life" (*Chuan Xi Lu*). There are many reasonable elements in the life-view of ancient China which are consistent with modernity and become important ideological and cultural resources of bioethics.

Bioethics in Asia must have its own characteristics and should make a special contribution to the bioethics of the whole world. It is a significant task to make use of the wisdom of the life and death viewpoint in Chinese traditional culture and give it new understanding, interpretation and development.

Shanghai Academy of Social Sciences
Shanghai, China

REFERENCES

Chinese Classics:

Analects of Confucius
Huang Di Nei Jing (*Interior Classics of Yellow Emperor*)
Lao Zi
Lie Zi
Meng Zi
Mo Zi
The Pictures of Twenty Four Filial Sons
Shi Wen (*Ten Questions*)
Sun S-M: *Qian Jin Yao Fang* (*Golden Prescriptions*)
Wang, Y-M: *Chuan Xi Lu* (*Records for Teaching and Learning*)
Xiao Jing (*The Book of Filial Piety*)
Xun Zi
Yi Zhuang (*A Commentaries on the Book of Change*)
Zhang Zai: *Zheng Meng* (*Book of Qian Cheng*)
Zhuang Zi

CHAPTER 14

PING DONG AND XIAO-YAN WANG

LIFE, DEATH AND END-OF-LIFE CARE: TAOIST PERSPECTIVE

In Chinese traditional philosophy, Taoists advocate "following nature" (*fa zi ran*) and claim that "coming into life and going out of death" (*chu sheng ru si*) are nothing more than a natural change. The view of considering life and death as day and night, waking and dreaming, provides an intellectual foundation for end-of-life care. In this paper we argue that the acceptance of the Taoist perspective on life and death will help dying people reach the end of life's journey and come up to the expectation of, and pursuit for, a good end with peace and dignity.

1. GRIEF OF THE DYING: PERPLEXITIES ON LIFE AND DEATH IN ANXIETY ABOUT DEATH

People think of life a lot, experience life a lot, and expect a lot from life, while they dislike talking about death, even dislike thinking of it, and especially fear death befalling them or others suddenly. To human beings, death is the most mysterious, horrible, and shocking thing that they have never experienced but know is unavoidable. It is natural for people to love life and fear death. To a dying person, the biggest tragedy is to immerse her/him in anxiety about death. Therefore end-of-life care aims at alleviating psychological and spiritual suffering.

In the book *On Death and Dying,* E. Kübler-Ross describes that the process of psychological reaction of patient who suffer from incurable diseases from learning terminal condition can be divided into 5 stages: denial, anger/indignation, bargaining, depression and acceptance (Kübler-Ross, 1969). Recently, some Chinese scholars, such as M-X Zhao, have observed a similar pattern of psychological reaction in more than 1000 cancer patients in the late stage of the disease (Zhao, 1992). However, D-Y Jiang *et al.* pointed out that the stages of psychological reaction of terminal Chinese patients are not the same as those identified by Kübler-Ross. They observed that before the stage of negation, there is a stage of avoidance in the reaction of Chinese patients (Jiang *et al.*, 1992). No matter how different their

R.-Z. Qiu (Ed.), Bioethics: Asian Perspectives, 147-155.
© 2003 *Chinese and International Philosophy of Medicine* 1:1 (February 1998): 107-120.

observations are, they converge on the same restless emotions of grief, suffering and anxiety once patients know the truth of their impending death.

In many cultures the topic of death has been dodged. Death can be expressed tacitly in English; it is said that in English there seem to be more than 50 ways to express "he/she died", and in Chinese there are even more. Chinese people can use different expressions to refer "death" or "die" in different occasions, such as *jiabeng* (the passing away of an emperor), *yunmi* (disappearance like a meteorite), *shishi* (passing away of a human life), *guoshi* (human life "going over"), *changbie* (leaving forever), *changli* (separating forever), *xianshi* (going away like a celestial being), *xisheng* (sacrifice), *xunnan* (giving one's life for a just cause), *juanqu* (laying down one' life), etc. There are nearly 100 words that express death. Both common people and notables see death as a taboo subject. In *Analects of Confucius*, it was recorded that Confucius did not talk much about death. Once when his student asked him about death, Confucius answered: "Without knowing life, how can you understand death?" (*Analects of Confucius: Xianjin*).

The taboo of death shows that people fear death and are concerned with *this life*. Although life and death are two opposite ends in human life, the views on them are interrelated. The view on death often exerts influence on the view of life and the view of life determines the attitude towards death. There are various reasons for people to fear death: some people fear the torment of dying; some fear where they might go after death; some hate to lose the joy of this world; and some regret that there are still unfulfilled duties. All these come from the perplexities about death.

A survey on Chinese people's attitudes towards life and death shows that the things that people who are faced with death are mostly concerned about are as follows (Cui, 1992):

Groups	Nothing exper- ienced	Don't know what will become	Don't know what will happen	Duties for family not fulfilled	Friends will be sad	All plans will be ended	Possible suffering	Other
Urban residents	366 15.93%	176 8.06%	206 8.97%	493 21.46%	231 10.06%	296 12.89%	471 18.15%	83 3.57%
Rural residents	34 16.11%	17 8.06%	19 9%	76 36.02%	17 8.06%	26 12.32%	17 8.06%	5 2.37%
Urban Catholics	20 6.92%	2 0.69%	88 30.45%	66 22.8%	77 26.64%	16 5.54%	4 1.38%	13 4.5%
Rural Catholics	9 2.25%	0 0%	275 68.75%	40 10%	10 2.5%	30 7.5%	0 0%	31 7.75%
University students	249 23.65%	21 1.99%	94 8.92%	207 19.66%	146 13.87%	167 15.86%	90 8.55%	54 5.13%

The reasons for fearing death and feeling anxiety can be divided into four kinds: (1) fearing where one might go after death; (2) fearing torment of dying and grieving of one's family members; (3) hating to leave one's enjoyable life, and (4) feeling depressed because of unfulfilled duties to one's family members. All these reasons are related to the viewpoint of human life. In the two groups of Catholics, the first reason is "Don't know what will happen after death". Influenced by Catholicism, Chinese Catholics are more concerned with what will happen after death. Their response is different from that of those Chinese who still advocate Chinese traditional culture, which is characterized by *zhensheng dansi* (treasuring life and regarding death with indifference), because the response of the two groups of Catholics is tainted with the color of Western religion. In the response of two groups of residents, the first reason is "regretting that there are unfulfilled duties for family members", and university students take "no longer experiencing anything" as the first reason. Those responses reflect the pragmatic feature of Chinese traditional culture with its focus on *this life*.

Chinese culture is characterized with treasuring life, pragmatism and focus on *this life*. The majority of Chinese don't believe in the existence of a soul that still exists after death. Once Confucius said that we offer a sacrifice to gods as if they do exist. Some Chinese worship spirits or Buddha in order to hope for living better by making a fortune and having good luck. Taoism is a Chinese native religion whose practitioners hope to live forever. Grown in the rich soil of Northern China and originating from the region of Qi and Lu, Confucianism occupied a central position in Chinese culture. Its emphasis on life *(shengsheng)* and its sorrow for death *(aisi)* philosophy has exerted deep influence on Chinese minds and their attitude towards their lives: focus on this world and pursue a happy life without considering death. When one lives, he/she cherishes and enjoys life; when life cannot be continued, one must wait for the coming of death. It is natural that in such a cultural atmosphere, people are frightened and anxious as death knocks on the door of life. We will argue that the Taoist conception of life and death would be a key to removing perplexities about death.

2. THE TAOIST CONCEPTION OF LIFE AND DEATH: NON-ACTION WHEN FACING NATURE

The core value of Taoist philosophical reflection on life and death is the worship of nature. Taoism emerged in Chu (now mainly Hubei and Hunan provinces), a powerful kingdom during the period of Springs and Autumns and Warring States, and it might be influenced by Chu's beautiful landscape rich with green hills and clear waters *(shan ming shui xiu)*. Chu culture is distinct from the mainstream of Chinese culture (Central China or *Zhongyuan* culture). The charm of Chu's culture shaped Taoist characteristics, such as advocacy of nature and simplicity, and an apathetic or free and easy attitude towards life and death.

The core concept and principle of Taoism is *tao*, which has a double meaning: *tao* as the origin of the universe and *tao* as the law of nature. As the origin of universe, *tao* is the first principle of all things, and it is transcendental, undefinable, creative, and generative. "Tao produced One. The One produced two. The two produced three. And the three produced the ten thousand things. The ten thousand things carry the *yin* and embrace the *yang*, and through the blending of the material force (*ch'i* or *qi*), they achieve harmony" (*Laozi*: 42; Chan, p. 160). Here One, two, three symbolize the process of creating all things between heaven and earth by *tao*. At the stage of One, *tao* was nothing (*wu*), there was no physical thing existing in the universe, but an undifferentiated mass of chaotic metaphysical psycho-physical entity called *qi*. At the stage of two, a process from nothing to being, *qi* differentiated into *yin* and *yang*. And at the stage of three, all things in the universe were generated from the interaction between *yin* and *yang*. As the first principle of the universe that generates everything in it, *tao* is shapeless, actionless, emotionless and will-less. *Tao* is not like gods or spirits in any sense. It is not a personified God who dictates the universe. *Tao* is the law of nature without intention, purpose and will. *Tao* generates all things and exists in them. *Tao* does not order anyone or anything, but the existence and evolution of them depend on it. Just as Zhuang Zi said: "The sky will not be so high if it does not follow *tao*; the earth will not be so vast and immense if it does not follow *tao*; the sun and the moon will not move if they do not follow *tao*; and things in the world will not prosper if they do not follow *tao*. This is the way tao works" (*Zhuangzi: Zhibeiyou*).

Although *tao* is the creator of the universe, the nature of *tao* is natural. Taoists claimed that "Man models himself after Earth. Earth models itself after Heaven. Heaven models itself after Tao. And Tao models itself after nature" (*Laozi*: 25; Chan, p. 153). Heaven, earth and human beings all model themselves after *tao* and *tao* models itself after nature. However, it is not to say that nature places itself over *tao*, overrides or manipulates *tao*, but rather, means the nature of *tao* is natural. A state that is called natural is one without human interference. *Tao* moves according to its own intrinsic trend. Everything, including the life and death of all creatures, follows the natural order of *tao*. Action that human beings take in society should take nature as the model. The Taoist's claim that "*tao* models itself after nature" or "heavenly *tao* is natural by achieving without acting" denies that nature would follow human will. On the contrary, the action that human beings take should follow the model of nature, and human *tao* should take heavenly *tao* as its model—non-action and natural.

Putting forward the proposition "*Tao* models itself after nature", however, Taoists do not lead people to escape from this world. In fact, what they claim is that human beings should not be arrogant or pretend to be omnipotent when faced with nature. Instead, they should go beyond the dichotomy between human and nature, and reach the harmony between human and nature (*tian ren he yi*) by means of the concept of *tao*. Human action should be guided by taking "nature" as the norm. The Taoist conception of *tao* is to explore the meaning of human life from a cosmological perspective.

It follows from the Taoist conception that life and death should model themselves after nature. For Taoists, "coming into life and going out of death" (*chu sheng ru si*) are just a change of natural order. From the viewpoint of Taoism, life and death are different only in appearance. Indeed, the difference between them is rather in the form of change. Both of them are transient forms of *qi*, which is naturally creative and transformative. "Death", as the loss of human life, may come too rapidly and too suddenly to an individual, but it is an unchanged phenomenon in the universe. Under heaven there is nothing that is born and grows but does not die. There is an appropriate time for people to come to this world and go to another world. Life and death are just like the change of day and night. Without life there is no death, and *vice versa*. Life and death alternate with each other. One comes, another goes. The course from life to death and from death to life is just like the change of four seasons: nobody can avoid it or change it. So for Taoists, human death is like sleeping peacefully between heaven and earth: for the living and the dead it is unnecessary to worry or fret about death.

From the microscopic or individualistic perspective there is birth, grow, and death of its own specific form. But from the macroscopic or cosmological perspective, anyone can turn into another by one form withering away and another form regenerating. All things in this world transform each other, and the end is the same as the beginning. Things change, and the beginning of change is linked to its end; all are interrelated with one another, and repeat themselves in an endless circle. Taoists name this *tian jun* (heavenly equilibrium). Under *tian jun*, the life and death of a human being is like an endless chain, and individual life or death is only one link. Knowing the truth, there will be no fear in death. Taoists hold that people should accept life and death naturally. Human life is like a flash of lightning—if you don't know the truth, you will see death as a deep abyss seen from a steep cliff. People should not confine themselves to the outer form of their body, only incorporating themselves into the mainstream of life. In the universe it would be possible for them properly arrange their lives and detach from the world of mortals.

Taoists emphasize that people should conform to the nature of all things that they deal with and take non-action when facing nature. In one sense, non-action means that we let nature take its course, and in another sense, it means that we should not take action unnaturally or against the nature of the thing we deal with. Nature "does nothing and everything is done." According to the Taoist view, there is an opposition between natural and artificial. Artificial is not natural, because artificial is what humans have made. However, the human is natural in her/his nature and origin. So humans should keep their own natures and their actions should conform to nature or the natural order. To conform to nature, you must take non-action. Zhuangzi said: "Only non-action is natural" (*Zhuanzi: Tianzifang*), "'Do nothing' and 'let nature take its course'" (*Zhuanzi: Shanxing*). Taoists especially oppose actions or activities that are done against nature and the nature of a human being, and emphasize the pursuit of a spiritual sphere of being in a natural state. Once reaching this spiritual sphere, one will not do things unnaturally. On the Taoist view, pursuing this spiritual sphere requires non-action, non-action requires no intention, no intention requires being oblivious to oneself, and being oblivious to oneself, one will enter into the natural state in which one will not care which is oneself and which is other. Nothing is unnatural in this spiritual sphere of being oblivious of oneself and others.

The Taoist philosophy of taking nature as a model and seeking spiritual freedom not only has exerted profound influence on Chinese traditional culture but also would give modern people some enlightenment on the issue of end-of-life care.

3. THE TAOIST IMPLICATIONS FOR THE END-OF-LIFE CARE

Issues in end-of-life care include: management of pain and suffering, psychological and spiritual care, family and social care, cardiopulmonary resuscitation, medical futility, withholding and withdrawing treatment, physician-assisted suicide, euthanasia and decision-making. Pain caused by disease can be controlled with the use of pain-killers, but suffering is a different issue. The difference between them lies in the fact that pain is largely physical, but suffering is spiritual, metaphysical or philosophical. Suffering is also hard to define, measure and express. However, a few people who have been back from the edge of death report that at the moment of death they experienced a kind of extraordinary calm and peace without any suffering. One person said:

> It seemed that my life was being hung between my two lips. I kept to the conventional way of closing my eyes; it looked as if this was conducive to pushing life out and I wallowed idly, willingly to let myself go. That thought just floated on the surface of my

soul as weak as anything else did. Only those people who let themselves slip into sleep
experience this feeling of which there is no sadness but mixed with happiness. I believe
that many people have experienced the same state (Lewis, 1966, p. 81).

If this reported experience of death is true, the torment of death is mainly
concentrated at the stage of dying before death.

Grief and anxiety are common symptoms of dying patients. So the key factor
in end-of-life care should aim at relieving the suffering souls of dying people. A
Chinese adage says "Mental worries should be cared for by mental drugs". The good
method to get rid of suffering relies on the detached theory inherent in the Taoist
conception of life and death. It would help people to go beyond the dichotomy
between "I" and "other", neglect the utility of human life, take a detached attitude
towards their own life and death, and get out of the mindset of being tenaciously
attached to the present life.

On the Taoist view, the feeling of fearing death comes from the dichotomy
between I and others, and the fantasy of trying to surpass nature to achieve the ideal
of living forever. Humans cannot change the natural order of life and death.
Therefore, living under the shadow of death, and facing the harsh reality of death
coming at any moment, it is natural for people to have the feeling of fear, grief and
sorrow regarding death. To break the dichotomy between life and death, people
should realize the "unity of life and death", gain the insight of the *tao* inherent in
external "things", "heaven" and "life", and integrate themselves and nature into
one—a unique spiritual sphere that common people can never experience. In this
spiritual sphere, people would surpass the limit of space and time, merge themselves
with the eternal stream of life in nature, and "co-exist with heaven and earth and
incorporate into one body with all things" in the universe.

Taoists claim that humans and things in the world are nothing but transient forms
of perpetual change in the universe, so the meaning of life should be explored in the
mental or spiritual world. People should shake off the bondage of various desires
and interpersonal entanglements so as to free themselves from mental torment about
loss and gain, enjoy the "absolute freedom" (*xiao yao*) of life, and enjoy complete
unrestraint by neglecting the dichotomy between life and death. Although Taoism is
abstruse, the Taoist naturalistic conception of life and death would help dying people
facing the harsh reality of death to transcend themselves, accept their tragic reality
with a detached attitude, and overcome anxiety and suffering preceding death.

The sense of historical duty fostered in the solid soil of the Confucian culture
overloads Chinese people's mentality. They work extremely hard all their lives,
shedding their heart's blood for their family, society, country and nation; after death,
they still will be possessed by this kind of duty. The purpose of building graves is to
bring benefit to future generations. This is praised as fallen flowers which "will
cultivate future flowers after being transformed into spring mud". The sense of duty
is important, which causes dying people to become immersed in fear, anxiety and
suffering about death. The Taoist view of life and death, which takes death as rest,
provides a theoretical foundation for dying people to receive psychological and
spiritual care.

Zhuang Zi said: "The great earth endows me with a physical form to dwell in,
makes me toil to sustain my life, gives me ease to idle away my old age, and offers
me a resting place when I die" (*Zhuangzi: Dazongshi*). The view of taking death as
rest is completely naturalistic. In the light of the above view, people who are faced

with affairs of human life have to accept natural things without differentiating them, and know the facts without making value judgements. Thing are just themselves, and so is life, death, and change. It should not be judged which is good or bad, beautiful or ugly, right or wrong, virtuous or vicious. If people want to do something without making value judgements, they have to be calm in their mind when accepting any change. On the contrary, once people make value judgements, they will bring about a variety of thoughts and actions so as to pursue one thing and avoid others. This constitutes the deep source of the human emotion of loving life and fearing death, a shackle to the true human spirit. So, the Creator (*zao hua zhe*) gives us life to toil, let us die to rest. People understand that it is the natural order for them to be born and to die, it is not necessary to be involved with individual value judgements and the emotion of loving life and fearing death.

Taoist scholars suggest common people should abandon the small wisdom of distinguishing right from wrong and merge themselves into one body with *tao* or nature. This is a kind of mysticism. Taoism guides people not to love life and to loathe death. However, they never persuade people to give up life and seek death. In fact, Taoist theory implies that after taking a detached attitude towards life and death, people can fulfil their allotted life span more lightly and make sorrow, anxiety and suffering disappear without a trace.

The anxiety of dying patients indicates that they have fallen into a cultural predicament in which the previously deposited traditional culture could not cope with the present harsh reality of death. At this moment, the Taoist view on life and death demonstrates a new outlook on life, with its characteristics of freshness, calmness and profoundness. It makes dying people no longer angry or grumbling, and helps them calm themselves down and face life and death with a detached attitude. Taoists respect the nature that is the same as a kind mother who comforts every painful heart with delicate love. From nature, people will get the heart of the universe. By eliminating value judgements of their own lives, the tragic mental complexes of dying people will disappear. Through a kind of free spirit of identifying oneself with *tao*, Taoists turn nature into something that eases the mind. Nurtured in the natural atmosphere of mystery, calmness and profoundness, people will lay down their anxiety about life and death, the pursuit of utility and various desires. They will gain understanding thoroughly from nature and gain true comfort to reach the state of ease inherent in "oneness of heaven and man" (*tian ren he yi*).

The Taoist view on "non-action facing nature" requires that people should respect nature, conform to the natural order, let nature take its course, and reject the attitude of seeking life and avoiding death. Lao Zi once said: "Three out of ten are companions of life. Three out of ten are companions of death. And three of ten in their lives lead from activity to death. And for what reason? Because of man's intensive striving for life" (*Laozi*: 50; Chan, p. 163). If the desire to seek life is too strong, it will do harm to life itself.

In clinics, dying patients with incurable disease go hither and thither to seek medical care hoping for a medical miracle. They, just like hungry people, are not choosy. In the course of treatment, if the condition is not improved or gets worse, the psychological overload caused by it will lead to the acceleration of death. It shows clearly that in end-of-life care, psychological and spiritual care is very important. By choosing non-action facing life and death, it does not mean that people will do nothing at all, but rather that people should do everything naturally or let nature take its course. This interpretation of non-action will help dying people to take a relaxed attitude towards a coming death.

In the preface of *Huang Ting Jing*, Ouyang Xiu (who lived in Northern Song Dynasty) said: "Tao is the way of nature. It is a natural principle for people to be born and to die. All sages from remote antiquity to nowadays agreed to nurture natural life with natural *tao*, and not to take action to shorten their natural life so as to fulfil one's own alloted life span" (see Ge, p. 50).

Taking nature as the model, Taoists do not advocate reckless human interference with natural life and death. According to that view, passive euthanasia, or withholding and withdrawing treatment, is appropriate in end-of-life care. Taoists believe that only those people who do not cling to life can really surpass the dichotomy between life and death. Those who always take life as the most important thing may often be frustrated in their process of dying. According to what was reported by one older lady, the only painful and depressing thing for her was to be disturbed in the course of death. Several times, she was given treatment to ensure oxygen supply or restore the body fluid and electrolytes, but every time she felt it was a kind of torment to be awaked, and she utterly hated the course of her death to be interrupted (Lewis, p. 44). The profound meaning of the Taoist view on life and death is to let people understand that life or death is a specific stage of natural evolution. Therefore, humans should conform to the natural order and not try to change it. Guided by the view of clinging to life, actions will be reckless, result in failure and lead to more suffering. When letting nature take its course, dying people will pass away with dignity in peace. This is a good end.

The purpose of the Taoist claim of "non-action" is to teach people to conform to nature and not to act recklessly. So active euthanasia or physician-assisted suicide is incompatible with Taoism. Both Lao Zi and Zhuang Zi believed that life and death are the change of the great *tao*. The value of human life does not lie in myself or others, because it is one form of great *tao*. So people should not grieve at death, especially not sorrow over life to such an extent as to give up life and seek death. People should live their "natural life span" and not come to a premature end. "Natural life span" is the time limit of life span entrusted by nature, while "premature end" of life is caused by external factors. On the Taoist view, everyone should try to experience the great *tao*, identify her/himself with it and adopt a detached attitude towards life: Let nature take its course.

The Taoist conception of life and death widens people's vision and helps them to merge their spirit with the great *tao* of the boundless universe. A secret key will be given to dying people and can be used to open their heart and eliminate or reduce suffering during their dying process.

Capital Medical University
Beijing, China

ACKNOWLEDGMENT

The original Chinese version of this paper appears in the thematic issue entitled 'Euthanasia' of *Chinese and International Philosophy of Medicine* 1: 1 (February 1998): 107-120; issue editor: Ruiping Fan. The authors wish to thank the journal for permitting translating the paper into English for publication in this volume.

REFERENCES

Chinese Classics:
Analects of Confucius . See *Selected Readings on Chinese Philosophy*, Book Pre-Qin Dynasty. Beijing: Chinese Bookstore.
Laozi. See *Selected Readings on Chinese Philosophy*, Book Pre-Qin Dynasty. Beijing: Chinese Bookstore.
Zhuangzi. See *Selected Readings on Chinese Philosophy*, Book Pre-Qin Dynasty. Beijing: Chinese Bookstore.

Modern Publications:
Chan, W.-T. (1973). *A Source Book in Chinese Philosophy*. Princeton, New Jersey: Princeton University Press.
Cui, Y.-T. (1992). 'Chinese attitude towards death and dying,' *Proceedings of the First International East-West Symposium on End-of-Life Care*, Tianjin.
Ge, R.-J. (1991). *Taoist Culture and Modern Civilization*. Beijing: Chinese People's University Press.
Jiao, D.-Y., & Cui, Y.-T. (1992). 'Psychological stages of dying patients and their treatment,' *Proceedings of   the First International East-West Symposium on End-of- Life Care*, Tianjin.
Kübler-Ross, E. (1969). *On Death and Dying*. London: Macmillan.
Lewis, T. (1996). *Notes of a Biologist Observer*, S.-M. Liu (Trans.). Changsha, Hunan: Human Press of Science & Technology.
Zhao, M.-X. (1992). 'On care for patients in end-of-life care,' *Proceedings of the First International East-West Symposium on End-of-Life Care*, Tianjin.

CHAPTER 15

DA-PU SHI AND LIN YU

EUTHANASIA SHOULD BE LEGALIZED IN CHINA: PERSONAL PERSPECTIVE

It can't be ignored that euthanasia is an issue which has social implications beyond individual discretion. Its practice is conditioned by various factors, including social, economic, cultural, ideological, ethical and legal factors. So far there is no consensus reached by all social and cultural groups or moral communities in today's China. Although many ethicists argue that euthanasia can be morally justified under certain strict conditions, according to the interpretation of China's criminal law, active euthanasia is illegal, so it can't be legally put into practice in clinical work. However, euthanasia, including active euthanasia, does exist. From our point of view, it is urgent to legalize euthanasia and it has been confirmed by innumerable, multileveled surveys and typical cases. Since the first Chinese legal case of active euthanasia in Hanzhong City in Shaanxi Province (1986) and a hotly debated lawsuit in Funing County, Jiangsu Province (1994), it can be seen clearly that the attitudes of Chinese people towards euthanasia have changed.

For example, in a survey of 400 people from different occupations and with different educational backgrounds, 80.13% of respondents advocated euthanasia (Shi, D-P, 1993). According to the data by Wang, Y-X (Shi, R., 1997), among the 563 terminally ill patients of a hospital in Shanghai in 1983, 1984 and 1985, 26% of the patients and their family members requested the withdrawal of life-saving treatment. There is evidence to show the rate is increasing. This essay is an attempt to argue from analysis of lawsuits, particular cases and survey results that euthanasia should be legalized in China. Here "passive euthanasia" refers to giving up, withholding or withdrawing life-sustaining treatment to let patients die naturally. "Active euthanasia" stands for intentionally causing terminally ill patients to die in a painless way in order to alleviate their suffering.

R.-Z. Qiu (Ed.), Bioethics: Asian Perspectives, 157-169.
© 2003 Chinese and International Philosophy of Medicine 1:1 (February 1998): 159-175.

1. TYPICAL LEGAL CASES

1.1. Typical Case 1

The first Chinese legal case of active euthanasia occurred at city of Hanzhong in Shaanxi Province (Yu *et al.*, 1989). Defendant Lian-Sheng Pu was a 46-year-old male doctor and director of the Hepatitis Section, Inpatient Department, Hanzhong Hospital for Infectious Disease. The other two defendants are Ming-Cheng Wang (male, 36 years old, with a junior middle school education background, staff member of sales section of Shaanxi No. 3 Printing and Dyeing Factory) and his younger sister Xiao-Ling Wang. On June 28, 1986, Pu was repeatedly implored by the Wangs to practice euthanasia on their mother Su-Wen Xia. Xia was a 59 year-old female patient who suffered late-stage cirrhosis, ascites, advanced hepatic encephalopathy, and severe ulcerative bed sores with unbearable pain. After these two defendants had signed on the prescription saying that "the family members request euthanasia", Pu prescribed 100 milligrams of compound chlorpromazine for the patient. The nurse only injected 75 milligrams into her. Six hours later, the fourth defendant, doctor Hai-Hua Li, who came on duty after Pu, after being enjoined by Pu and requested again by the Wangs, prescribed another 100 milligrams of compound chlorpromazine for the patient. Gui-Lin Zhao, a nurse on duty, injected it into the patient. Xia died in the emergency treatment room of Hanzhong Hospital for Infectious Diseases at 5 a.m. June 29, 1986. On June 30, Xia's four children buried their mother.

But later the patient's other two daughters asked the hospital to compensate for their mother's death that they thought was caused by the injection of chlorpromazine. Her eldest daughter, Jian-Min Wang, was an accountant at the Vehicle's section of the No. 9 Mailbox of Hanzhong City. Another elder daughter, Xiao-De Wang, was a worker of the No. 3 Bureau of the Ministry of Hydroelectricity. On July 3, 1986, they sued Pu for intentional murder to the Public Security Bureau and the Procuratorate of Hanzhong City. On July 4, four medical personnel brought a charge against Pu to the Precuratorate of Hanzhong City as well. The Public Security Bureau of Hanzhong City investigated the case and on September 20, 1986, detained Pu, Li, Xia's son and youngest daughter for intentionally killing. After M-C Wang and X-L Wang were detained, the plaintiffs, J-M Wang and X-D Wang, repented and asked that the charges be withdrawn. But the Public Security Bureau and the Procuratorate of Hanzhong City refused. On December 20 in the same year, the four defendants were released upon bail pending trial by the Public Security Bureau.

On March 31, 1987, the Evaluation Committee for Malpractice of Hanzhong Prefecture investigated the case and believed that Xia's death was caused both by the disease and the effect of chlorpromazine, and that the injection of the drug hastened her death. According to this, the Public Security Bureau of Hanzhong City reported to the Procuratorate of Hanzhong City that the four defendants should be arrested. In September 1987, the Procuratorate approved the arrest of Pu and M-C Wang for intentionally killing. On February 2, 1988, the case was submitted to the People's Court of Hanzhong City. Li and X-D Wang were given immunity from prosecution.

On June 23, 1988, the People's Court of Hanzhong City released Pu and M-C Wang on bail pending trial. Zhan-Nin Zhang, the lawyer of the defendants, challenged that Lei, the Director of the Hospital for Infectious Disease and a member of the Evaluation Committee for Malpractice of Hanzhong Prefecture, had a grudge against the defendant Pu and thus should not be involved in the case.

Therefore, on September 3, 1988, the People's Court of Hanzhong City entrusted the Department of Forensic Medicine, Shaanxi Higher People's Court with the work of appraising the cause of death again. After the investigation, on February 22, 1990, the doctors in the Department of Forensic Medicine confirmed the Xia's diagnosis of "having contracted ascites due to cirrhosis, hepatic dysfunction, hepatic encephalopathy, exudative ulcerative bed sore of II to III degree" made by Hanzhong Hospital for Infectious Disease. They also confirmed that the chance of the patient getting better was small. They found that the total amount of compound chlorpromazine injected into Xia's body twice was only 175 milligrams, with 87.5 milligrams of chlorpromazine and 87.5 milligrams of phenergan. So the injection amount was normal. Since the patient died 14 hours after the injection, the blood pressure didn't drop abruptly, and there was no respiratory center inhibition, the injection of chlorpromazine only deepened the comatose state and was not the key cause of death. The key causes were cirrhosis and hepatic encephalopathy. The same conclusion was drawn by the experts from the Evaluation Committee for Malpractice of Shaanxi Province.

In April, 1991 the People's Court of Hanzhong City judged Pu and M-C Wang not guilty at the first trial. Unsatisfied with the result, the Procuratorate of Hanzhong City appealed against the judgement, and thinking that some facts in the judgement were not true, Pu and M-C Wang appealed as well. Considering that it was intentional behavior but had only mildly harmful consequence on the society, the Shaanxi People's Higher Court regarded the judgement made by the People's Court of Hanzhong City correct and rejected the two appeals. Thus the 5-year lawsuit was ended at last.

Dispute at the Court: The public prosecutor insisted that Pu and M-C Wang should be sentenced to death in accordance with Article 132 of China's Criminal Law, i.e., they committed the crime of intentionally killing a human. But the defendants' lawyer believed that the behavior of the two people was not the direct cause for the death, and therefore, they should be judged not guilty.

There were three kinds of opinions on the case at trial. The first one was that this case should be judged in accordance with China's current Criminal Law and should not be influenced by the opinions of scholars, mass media or the public on euthanasia, for euthanasia has not been legalized in China and it would therefore be against the current law if anybody practiced euthanasia on others. However, because of the mildly harmful consequences to society, it could be treated as a minor crime, and the two defendants should be immune from criminal sanction in terms of Articles 132, 22, 23 and 32.

The second opinion was that the two defendants didn't commit a crime and thus should be judged not guilty in term of Article 10. The reason is that, euthanasia is beneficial to society, and its purpose is to relieve a terminally ill patient from unbearable pain or/and suffering. The two defendants' motive was to alleviate the pain and suffering of the patient, and their behavior was not the direct cause of death.

The third opinion was that the defendants of the case should be judged guilty in terms of Article 32, for they caused the patient's death under the cover of "euthanasia" and it was harmful to society. "Killing through euthanasia" was very similar to intentional murder, and should be treated as intentional murder by analogy as there has been no regulation in China's Criminal Law on "killing through euthanasia."

1.2. Typical Case 2

This legal vase of euthanasia happened in Funing County, Jiangsu Province (Shi *et al.*, 1994). Li Chen was a 30-year-old woman farmer with junior middle school education background in Tanzhuang Village, Yilin Town, Funing County, Jiangsu Province. On January 3, 1994, Chen killed her husband Haizhi Du, who was a 31-year-old suffering from advanced liver cancer, by tightening a rope around his neck. Li and Du were deeply in love, and Chen was a simple and honest woman. Before his death, Du Haizhi requested many times that Chen and the physician in charge help him to die because of the unbearable pain. Out of deep sympathy and love for her husband, Chen killed him. After Du's death, his brother' wife, Wei-Feng Zhou (an illiterate farmer, 27 years old, remarried after her divorce, on bad terms with Chen) found the trace of rope in the neck of the dead, so she questioned Chen, who admitted what she had done to her husband. Then, Zhou, together with another of Du's relatives, reported the case to the judicial department of Funing County. After collecting evidence and examining the body, the county court sentenced Chen to 3 years imprisonment, according to Article 132 of China's Criminal law.

Responses: The case was disclosed to mass media first by Zong-Lin Chen, secretary of Funing County Court, and caused a strong response. More than 60 press units across the country reported the case. On May 8, 1994, the Chinese Central TV Station (CCTV) made a follow-up report of the case, and attracted attention from lawyers and bioethicists. The issue of the legalization of euthanasia was raised again in the debate. Li Chen, bitterly repentant and perplexed, won the local people's deep understanding and sympathy due to her simple nature and her considerate care for her sick husband. Because Chen had to serve her sentence, nobody took care of her son, which garnered people's attention. Some bioethicists regarded the nature of the case as the crime of intentionally killing a human, though upon the request, but not as euthanasia. However, the case was not serious, and the measure of penalty was suitable. The public security department, procuratorate and court of the County had consensus on the treatment of the case, that she should be sentenced to the shortest imprisonment because of her non-evil motive and the controversy on euthanasia in society, though committing the crime of intentionally killing a human. This case made euthanasia known all over Funing County. Many people thought the penalty too heavy, and a number of other people held the view that she should not be punished at all. The debate on the possibility of legalizing euthanasia was stirred throughout the country again.

1.3. More Cases of Terminally Ill Cancer Patients

In 1989, China, Britain and the USA jointly ran a research program entitled "Investigation on the Causes of Death over 3 Years from 1986 to 1988", in which Xi'an was one of the spot-check cities. 37,225 people were investigated. The first four causes of death were malignant tumors, cerebrovascular diseases, heart diseases and injuries/poisonings. These accounted for 70.64% of the total deaths. Patients suffering from malignant tumors amounted to 118.2 per 100,000 people. Malignant tumor is now the number one cause of death, and was number two in the 1970s.

General Hospital of Shenyang Military Region concluded from 2,749 cases collected from 1981 to 1989 that the number of malignant-tumor-stricken patients was at the top, accounting for 27.8% of the total deaths and, among them, the largest segment is people between 56-65 years old. Death caused by malignant tumor was increasing year by year: the number in 1989 was 75% more than that in 1981 (Xing et al., 1991).

The aging problem facing China is unavoidable. There were 170,000,000 elderly aged over 65 in China in 2000, amounting for 13.6% of our population of 1,248,000,000. In the middle of the 21st century, the population of elderly will reach 300,000,000, 20% of the total population of 1,471,000,000. More and more elderly have expressed that they would not want to continue living in pain and suffering when they become terminally ill.

1.4. Particular Cases

In the journal *Chinese Medical Ethics*, there were collected and published almost one thousand cases from 1989 to 1995. 2,537 of these cases involved active or passive euthanasia, and 176 out of these 2,537 cases were active euthanasia. The following are some examples.

There was a case of active euthanasia in the Tumor Division of the No. 2 hospital affiliated with Henan Medical University, Henan Province (Xie *et al.*, 1990). The patient was at an advanced stage of cancer, and taken good care of by his family. Out of deep love to the patient, the family asked for euthanasia. The request was approved with formalities and kept confidential.

Seven cases collected in Nanchang, Jiangxi Province (Zhang *et al.*, 1994) were all active euthanasia of patients in advanced stages of cancer, who insistently requested euthanasia several times, but were rejected. They ended their unbearably painful lives by jumping from a building or by hanging themselves. Only one of them was rescued when committing suicide but later died from refusing medical treatment. Among those seven, four were male and three female. The oldest was 74 years old while the youngest was 42, with an average age of 51. They all had comparatively high educational backgrounds.

The Department of Internal Medicine of an army hospital in Shanghai (Zhao *et al.*, 1994) collected 3 cases. On was in the late stages of cancer, another had suffered brain trauma, and the third had hepatic encephalopathy. Two of the patients were female and 1 male; the youngest was 11 years old and the oldest was 57. One patient requested active euthanasia but was rejected and died passively. Another died of passive euthanasia. In the third case, active euthanasia was performed, but the performer was sued in court and the case caused legal dispute.

In the Division for Neonates of Shaanxi Hospital for Women and Children (Ni, 1993), there were 50 cases from January 1989 to March 1992 in which parents gave signed permission to discontinue treatment to a newborn. Among them, 26 were male babies and 24 female. Euthanasia was not because of sex selection, but because of some objective reason such as fatal disease or severe congenital malformation. 49 of them were of Han nationality, while one was of Hui nationality; 41 cases were passive euthanasia and 9 were active euthanasia. It was also shown in the cases that active euthanasia with merciful means had little influence on the surrounding baby patients, while passive euthanasia in the prolonged desperate struggle had great negative influence on them. It not only intensified the patients' pain, but also tormented the compassion of all concerned, parents, doctors and nurses.

Patients in the 19 cases collected in Nanjing, Jiangsu Province (Yan *et al.*, 1993) were all at the end stage of cancer or suffered prolonged illness without hope of recovery. They were extremely painful and all performed active euthanasia by themselves by means of taking poison.

Inner Mongolia Medical University (Li *et al.*, 1994) analyzed 613 patients over 60 years old, among which 139 patients (22.8%) committed suicide because of hopeless geriatric diseases and cancer at the terminal stage.

> These cases indicate that the practice of euthanasia in China has encountered many setbacks without laws to regulate the practice, but cruel reality forced people to adopt spontaneous methods for a "comfortable death" from time to time. Those terminally ill patients with fatal diseases ended their own lives by suicide when they could not endure the torment of pain and/or suffering and could not get others' help. The numbers of these patients cannot be neglected.

Euthanasia contradicts the notion held by many people that suicide is an immoral behavior of giving up and escaping from the responsibilities for society and family, and of bringing great sorrow to the family. This causes an ethical dilemma between advocating euthanasia and preventing patients in extreme pain from committing suicide. At present, when the legalization of euthanasia is not in the foreseeable future, can we allow those who meet the strict conditions for practicing euthanasia to commit suicide so as to relieve them of their empty or meaningless social responsibilities? Can this be a transition to euthanasia?

There have been some socio-cultural factors that have a negative impact upon the practice of euthanasia in China. Some traditional values, and the free medical service system, are the main reason why many doctors and patients' relatives and friends would rather keep patients in a state of unbearable pain than manage to relieve their pain and sufferings. Due to the influence of traditional notions and conventions such as filial piety, which promotes doing everything to keep parents alive even if the situation is hopeless or meaningless, and prohibits doing anything to shorten or hasten parents' deaths, people will not and dare not do what others are not doing. Many medical professionals think the faith of "Healing the wounded, rescuing the dying, practicing revolutionary humanitarianism" is in conflict with euthanasia. About 20% of the population in China enjoy labor insurance and public health service. The other 80% pay for medical treatment at their own expense. The dying patients who pay their medical costs at their own expense and their family's tend to accept euthanasia. However, those patients enjoying labor insurance and public health service and their families tend to hold a negative attitude towards it for they don't care about the scarcity of public resources. There are still a few people who are reluctant to care for their terminally ill parents among the supporters of euthanasia.

2. SURVEYS IN THE PUBLIC AND THEIR ANALYSIS

2.1. Survey 1

A survey including 3,197 citizens in urban and rural areas and Chinese Catholics was conducted by the End-of-Life Care Research Center, Tianjing Medical University (Ho, 1991). The result shows that religious belief has much influence on the outlook on death. To the question of "What is the most important thing you are concerned with when dying?", the most frequent response is "unfulfilled obligations to the family." It shows that Chinese have a very strong sense of responsibility to their family. Moreover, citizens in rural areas are more likely than Chinese Catholics or those in urban areas to prefer "to save the terminally ill if physicians possibly can."

2.2. Survey 2

A survey (Han *et al.*, 1992) on questions concerning death has been conducted in 454 Chinese individuals with different identities, ages and levels of education. The result shows that 65 percent of respondents favor the withdrawal of life-sustaining treatment from very severely deformed patients. 45.6 percent favor withdrawing it from seriously defective newborns. 27.5 percent of respondents favor practicing euthanasia on patients at terminal stages, despite their economic and political status. It also shows that those who oppose euthanasia are mainly the elderly and the less educated. Only 25.0 percent of respondents are willing to donate their organs after death, 21.7 percent of them think "good death is better than bad life," 7.2 percent of them think "bad life is better than good death." 57.5 percent of them are not afraid of death, in which 81.0 percent are elderly. 54.2 percent of respondents are willing to die without pain and suffering.

2.3. Survey 3

Statistics based on 664 responses to a questionnaire (Pi *et al.*, 1989) indicates that 588 people (88.6 percent) favor euthanasia, 76 (11.4 percent) oppose euthanasia. Among those who favor euthanasia, 9.4 percent accept active euthanasia. But among the supporters of euthanasia, 41.8 percent can accept euthanasia both in reason and emotion, while 58.2 percent can only accept in reason but not emotion. Among the opponents of euthanasia, 68.1 percent can't accept euthanasia mainly for emotional reasons.

2.4. Survey 4

In this survey 557 responses were received from 600 copies initially distributed (Fan *et al.*, 1995). The response rate is 92.8 percent. 54.0 percent of respondents support euthanasia, among whom 82.1 percent accept active euthanasia, and 17.9 percent accept passive euthanasia. 34.6 percent of respondents oppose euthanasia. 11.3 percent are neutral. Supposing that there is a terminally ill patient with an incurable disease who is suffering painfully, 55.8 percent respond that they accept euthanasia, 28.9 percent do not accept, 15.3 percent are neutral. The attitudes of people of different groups towards euthanasia are listed in the following table:

Attitudes of Different Groups of People Toward Euthanasia

Status	Respondents	Supporters		Opponents		Neutral	
	No.	No.	%	No.	%	No.	%
Urban	449	257	57.2	139	31.0	53	11.8
Rural	108	44	40.7	54	50.0	10	9.3
Male	296	156	52.7	119	40.2	21	7.1
Female	261	145	55.6	74	28.4	42	16.1
Primary school education	55	14	25.5	39	70.9	2	3.6
Middle school education	347	193	55.6	112	32.3	42	12.1
College education	155	94	60.6	42	27.1	19	12.3
No professional title	231	102	44.2	116	50.2	13	5.6
Junior professional title	226	137	60.6	52	23.0	37	16.4
Middle professional title	80	49	61.3	22	27.5	9	11.3
Senior professional title	20	13	65.0	3	15.0	4	20.0
Farmer	71	16	22.5	52	73.2	3	4.2
Worker	63	28	44.4	30	47.6	5	7.9
Judge	40	21	52.5	15	37.5	4	10.0
Administrative cadre	101	57	56.4	37	36.6	7	6.9
Medical professional	282	179	63.5	59	20.9	44	15.6
Common people	447	234	52.3	158	35.3	55	12.3
Bureau and department personnel	76	45	59.2	25	32.9	6	7.9

Regiment chief and higher rank officer	34	22	64.7	10	29.4	2	5.9
Youth League member	135	94	69.2	26	19.3	15	11.1
Chinese Communist Party member	207	121	58.5	63	30.4	23	11.1
Democratic party member	3	1	33.3	2	66.7	0	0
Non-party personage	212	85	40.1	102	48.1	25	11.8
Personnel with free medical service and labor protection	328	181	55.2	98	29.9	49	14.9
Personnel without free medical service and labor protection	229	120	52.4	95	41.5	14	6.1
Healthy person	489	262	53.6	168	34.4	59	12.1
Person with serious chronic disease	60	34	56.7	22	36.7	4	6.7
Person with incurable disease	8	5	62.2	3	37.5	0	0

2.5. Survey 5

The survey was conducted in 125 students with on-the-job training and 46 undergraduates of clinic medicine at a medical college (Li et al., 1994) Among the on-the-job training students, 115 support euthanasia (92 percent), 4 oppose (3.2 percent), and 6 are neutral (4.8 percent). Among undergraduates, 41 support euthanasia (89.3 percent), 2 oppose (4.35 percent), and 3 are neutral (6.25 percent). The total number of the two groups is 171, among whom 156 support (91.32 percent), 6 oppose (3.51 percent) and 9 are neutral (5.26 percent). The numbers of neutrals are a little bit more than opponents. The reason is that they can accept euthanasia in reason but not in emotion.

2.6. Survey 6

The survey was conducted on 317 medical doctors and nurses at junior, middle and senior levels (Fan et al., 1995). 291 copies of questionnaire were given back. The result is as follows:

The attitudes toward active and passive euthanasia are: 52 percent support both, 14.43 percent support active euthanasia, 8.2 percent oppose both, 16.8 percent are neutral, 7.6 percent think the decision of whether to practice euthanasia should be decided by the manager of the hospital. 72.9 percent think the prerequisite of euthanasia is the will to accept euthanasia by patient and his family.

Among the opponents, 31.6 percent think the practice of euthanasia will miss the opportunity of cure, 18.9 percent think accepting euthanasia is a kind of compromise to medical difficulties, which will prevent medical science from progress, and 8.6 percent think euthanasia is in conflict with a doctor's responsibility.

79.4 percent of respondents think that the main difficulty of performing euthanasia is the imperfect legal system, 54.3 percent of them think it is traditional values and beliefs, and 49.5 percent think it is the lack of education among medical workers or patient's family members.

This survey also indicates that the concept of euthanasia understood by 96.9 percent of medical workers is correct, while 3.4 percent identify it with "the right to hasten a patient's death."

64 percent have seen euthanasia in the clinical practice. More than 53 percent have known that medical doctors were requested to perform euthanasia by patient's family. 50 percent think a patient and her/his family have the right to refuse life-sustaining treatment.

These statistics show that most citizens without free medical service and a small portion of citizens with free medical service would choose to withhold or withdraw treatment, while most citizens with free medical service and some with good economic condition will desire receiving treatment until they die. But among 1.2 billion Chinese, only about 20 percent enjoy public health services and labor protection service. More than 80 percent do not enjoy free health service. Therefore, to forgo treatment is indeed a very common phenomenon, especially in rural areas. The survey of medical workers mentioned above has also demonstrated this.

3. EUTHANASIA SHOULD BE LEGALIZED

From these cases, surveys and real practices it is known that passive euthanasia exists in daily life, while cases of active euthanasia are few and silently practiced even though it is illegal.

Once accused, the suspect will be judged under the category of Article 132 of China's current Criminal Law ("the crime of intentionally killing a human"). Interestingly, all defendants have been sentenced only in the lightest sense of the crime of "killing", i.e., three years imprisonment. This is because the courts took into consideration the "good motivation" and "mild harm to society" that the behavior of the convicted brought.

We will argue that it is necessary for China to work out a specified law on euthanasia to make euthanasia legalized, and then the person who performs euthanasia would not be convicted as guilty of "intentional homicide" or "murder".

First, the subjects who receive euthanasia are terminally ill with incurable, fatal disease or deformity. They are either in irreversible coma or unbearable and intractable pain and/or suffering. Euthanasia is voluntarily and insistently requested by the patient her/himself or her/his family when he/she is incompetent. It is unfair to place such a special medical case under the crime of murder. The difference between euthanasia and murder lies in: (1) the agent of murder is motivated by her/his evil intention, but in cases of euthanasia the agent is motivated by the good intention of hoping to relieve a patient's unbearable pain and suffering; (2) in cases of murder the direct or fundamental cause of the victim's death is the murderer's action, but in cases of euthanasia the direct or fundamental cause of the patient's death is the disease. We also can use double effect to justify euthanasia: the direct and intended effect is to relieve a patient's pain and suffering, and death is the indirect and unintended effect of the action. The problem is that when the agent took the action of euthanasia, he/she knew that her/his action would bring the patient death. However, at this juncture, death is the only way to relieve her/his pain and suffering. The question is: Is it too cruel to see a human dying with unbearable pain and suffering against her/his sincere wish to die peacefully and her/his lifetime valuing of and belief in a good life and a good death?

Second, the humanitarian argument against the legalization of euthanasia is invalid. Some Chinese people argue that the legalization of euthanasia runs counter to humanitarianism. Humanitarianism in China is a paramount principle in medical practice. In their view, humanitarianism means that the value of man's life is a primary good higher than anything. It follows from their view that suicide is always immoral, so any kind of euthanasia, even for a terminally ill patient with unbearable and intractable pain and suffering, is also immoral. Ironically, many Chinese advocates of euthanasia also use the concept of humanitarianism to justify euthanasia. For them, the fundamental requirements of humanitarianism are to reduce human suffering, and respect the wishes of patients. They take euthanasia as a humanitarian choice in certain circumstances.

In fact, humanitarianism is an ambiguous concept, with different meanings to different people. What humanitarianism implies regarding the issue of euthanasia depends upon which particular theory of humanitarianism is accepted. Nobody can or should impose her/his own particular interpretation of humanitarianism on any other in a multi-cultural society with different moral communities. It follows that nobody has moral authority to coerce those who oppose euthanasia to accept or to be involved in any act of euthanasia. However, by the same token, opponents of euthanasia should not use the law to prohibit advocates of euthanasia from accepting or performing euthanasia. In our opinion, drafting a particular statute on euthanasia will create the opportunities that will better protect opponents from being involved in it involuntarily and ensure that advocates have access to it safely.

Third, as for the dispute on euthanasia, many people feel embarrassed that they support euthanasia in reason but can't accept it in emotion. It is natural to feel pain and difficulty in making such a choice regarding living and death. But what we need to do is to place elements of reason and elements of emotion in a moral balance to reach a moral conclusion. And we should always do the morally right deeds but not the morally wrong ones based on our moral intuitions of right and wrong both in reason and in emotion. The legalization of euthanasia may create a moral and legal opportunity for us to make a more humanitarian choice.

Fourth, many opponents of euthanasia are afraid that once euthanasia is legalized, it will be abused. Such thinking is reasonable. On the problem of human death, we should be very careful and cautious. However, like the possibility of abuse of any other law in the world, we shouldn't reject drafting a law because of fear of abuse. Not having a law doesn't mean there is no problem. In fact, it is misleading to put euthanasia cases under the category of Article 132 of current Criminal Law in China. Actually, it is more dangerous to perform euthanasia quietly in order to avoid accusation or punishment by the court. On the contrary, having a perfect statute on euthanasia will avoid and reduce the opportunities to abuse it in a large degree.

Finally, some people object to legalizing euthanasia in China because they are afraid that it will cause social instability. According to their view, given the reality that there are still a substantial number of Chinese people who object to euthanasia (although more than 50 percent of Chinese people support it), if euthanasia were legalized, these people would be upset and dissatisfied with the society and thereby social solidarity and stability would be undermined. However, we believe that this concern involves a confusion between two different attitudes. One attitude is that "I, as a person, do not permit myself to participate in any act of euthanasia because I believe it is morally wrong." The other is that "I will never permit anybody to be involved in euthanasia." The first attitude does not necessarily lead to objection to the legalization of euthanasia. Only the second attitude leads to objection to the legalization of euthanasia. In our opinion, most Chinese opponents of euthanasia in fact hold the first rather than the second attitude. Therefore, the worry of causing social instability due to legalizing euthanasia is in fact groundless.

Xi'an Medical University
Xi'an, Shaanxi Province, China

ACKNOWLEDGMENT

The original Chinese version of this paper appears in the thematic issue entitled 'Euthanasia' of *Chinese and International Philosophy of Medicine* 1: 1 (February 1998): 159-175, issue editor: Ruiping Fan. The authors wish to thank the journal for permitting translating the paper into English for publication in this volume.

REFERENCES

Fan, Y-N *et al.*: 1995, 'An analysis of responses of medical professionals to a questionnaire on euthanasia', *Chinese Medical Ethics* no. 3, 46.

Han, J-D *et al.*: 1992, 'A preliminary approach to the attitude of citizens towards Death', *Chinese Medical Ethics* no. 2: 50.

Ho, Y-C: 1991, 'Chinese attitude towards death', *Chinese Medical Ethics* no. 4, 60.

Li, Z-T *et al*: 1994, 'Perception of medical students to euthanasia', *Chinese Medical Ethics* no.5, 52.

Ni, X-M: 1993, '50 cases of forgoing treatment from neonates and euthanasia',*Chinese Medical Ethics* no. 2, 55.

Pi, I. *et al.*: 1989, 'A sociological and psychological survey on euthanasia', *Chinese Medical Ethics* no. 1, 48.

Shi, D-P: 1993, 'On euthanasia and its cultural background', in D-P Shi, *Ethical Issues in Medicine*, Xi'an: Northwest University Press, p. 23.

Shi, D-P, Yu, L. *et al.* (eds.): 1994, 'Fault of love', a part of TV series on euthanasia, produced Xi'an Medical University.

Shi, R.: 1997, 'An interpretation of euthanasia', in R.Shi, *Collected Papers*, Shaanxi Normal University Press, p. 116.

Xie, B-L, Zhang, Y-J: 1990, 'Comments on five cases of euthanasia', *Chinese Medical Ethics*, no. 1, 54.

Xing, R-J, Zhou, Q-G: 1991, 'The trends of medical ethics from the viewpoint of rhe change of death caused diseases and ages', *Chinese Medical Ethics* no. 5, 60.

Yan, L-F, Sun, Z-Y: 1993, 'Medical and ethical analysis of 171 cases of suicide', *Chinese Medical Ethics* no. 1, 54.

Yu, L., Shi, D-P *et al.* (eds.): 1989, 'A deep call', a part of TV series on euthanasia, produced by Northwest College of Political Science.

Zhang, C-N *et al.*: 1994, 'The necessity and urgency of legalizing euthanasia manifested in the eight suicide case of terminally ill', *Chinese Medical Ethics* no. 1, 64.

Zhao, L-L, Kang, Y-H: 1994, 'On euthanasia', *Chinese Medical Ethics* no. 2: 44.

PART IV:

BIOETHICS, POLICY, AND LAW IN ASIA

CHAPTER 16

RE-FENG TANG

CHINESE POPULATION POLICY:
GOOD CHOICE *VS.* RIGHT CHOICE

The paper is not intended to argue that Chinese population policy is good or bad, right or wrong. Instead, it argues that correct thinking on population policy should concentrate on goodness instead of rightness. The first part of the paper points out that Chinese thinking on population policy concentrates on goodness while Western thinking concentrates on rightness. In the second part, I ascribe these two ways of thinking to two kinds of ethics, that is, the ethics of good and the ethics of right, and conclude on the basis of Gilligan's feminist ethics and Dreyfus' phenomenological studies in moral development that "good" ethics is more mature than "right" ethics. In the third part of the paper, I shall analyze the implications of "good" ethics on population policy.

There are two kinds of approach to population policy. One is the scientific approach, which addresses the effects of population growth and population policy on every individual and society as a whole. The other is the philosophical approach, which has little to do with the analysis and prediction of facts, but only provides a methodological foundation. That is, it tells us how to decide if a certain kind of population policy should be accepted or not. It is the latter which will be explored in this paper. In what follows I will not show whether the current Chinese population policy is accepted or not or to what extent it is accepted. Instead, what I will do is analyze the ethical thinking behind the population policy and some criticisms of it. I will argue that "good" ethics is a mature ethics while "right" ethics is less mature. My conclusion is, Chinese thinking on population policy is a kind of mature "good" ethics which we shall carry out thoroughly. That is, we shall always consider whether our actions are good at every stage or level; we should not be satisfied with just doing the "right" thing.

R.-Z. Qiu (Ed.), Bioethics: Asian Perspectives, 173-183.
© *2003 Kluwer Academic Publishers. Printed in the Netherlands.*

1. CHINESE POPULATION POLICY: THE CURRENT SITUATION
AND ITS CRITICS

Chinese population policy is a kind of birth control policy. The policy is based on the thought that population control is good for Chinese people. According to this way of thinking, whether a policy is desirable or not depends on whether or not it is good for the subjects concerned. For those who support the birth control policy, the reason is very simple: birth control policy is good for Chinese people. It is not only good for Chinese people as a whole, but also good for every Chinese as an individual, so it is a good choice and right choice. In Chinese thinking, good is enough; a good choice is sufficient to be considered a right choice. This is an experience-directed thinking or a matter-of-fact attitude. We can see that those movies (such as *The Sweet Career*) which aimed to help people to understand the population policy always tried to show that a family with too many children would get into a mess in every respect; no one could enjoy her life properly. And opinions against the birth control policy were also based on the same way of thinking, although each side assesses the facts differently. At that time, people thought, "More people, more strength"; "More people can collect more firewood and make a brighter light", or in more sexist way, "More sons, more fortune". Although with different conclusions, they are all based on exactly the same way of thinking.

Recently, Chinese population policy has faced a different kind of criticism, that birth control policy is a violation of one of the fundamental human rights—the right to reproduce. The evident feature of this kind of critique is that it says nothing about the goodness of the policy, but only emphasizes that the policy violates some important ethical principle. Their arguments mainly consist of two aspects.

One aspect focuses on the universality of ethical principles. They argue that Chinese birth control policy is based on the Chinese traditional culture which says that the interests of an individual should be subordinated to the interests of the whole society. And for these critics, the interests of any individual should not be interfered with; thus the traditional cultural attitude is not desirable. But we can see from what I have just said that neither kind of Chinese population policy has ever emphasized the distinction between the interests of an individual and the interests of the whole society, and there has been no encouragement for individuals to sacrifice themselves to the whole society. The main point is just to suggest that everyone should make a wise choice for herself. More specifically, the supporters of the birth control policy are actually against the Chinese traditional culture which says "More children, more happiness" and "Have a son for your old days," etc. So, it is obviously implausible to argue that the birth control policy is based on any kind of Chinese traditional culture.

For some Western critics, population policy is not a problem of good or bad, but a problem of right or wrong. According to this way of thinking, rightness could be independent of goodness. For them, the crucial question is: Is it right to limit the right of reproduction? The question of good or bad is not to be considered. This is principle-directed thinking. We can see that this is a straightforward and simple one-word argument: it is wrong because it is not right. Once we think in this way, there is almost no way to argue against it. But is this the right way of thinking? What shall we do when right things contradict each other? Can one thing be more right than the other?

To be honest, this way of thinking does conflict with typical Chinese thinking. There is an old Chinese story named "Zheng ren mai lü" (A man from Zheng Kingdom sold shoes) which criticizes exactly this former way of thinking. Once there was a man who wanted to buy a pair of shoes for himself. Before he went to the market, he measured his feet to get the linear measure. Then he put the linear measure on his seat. When he went to the market and got a pair of shoes, he found that he had forgotten to bring the linear measure with him. So he said: "I have forgotten to bring the measure with me." And then he went home to fetch the measure. When he was back again, the market was closed, so he could not buy his shoes. Someone asked him: "Why not try the shoes with your own feet?" The man said: "I believe in the linear measure rather than my own feet." After that, the story has been used to criticize those people who stay with principles mechanically without knowing how to be flexible with real situations (*Han Fei Zi; Huai Nan Zi*).

The story lends itself to a thought experiment. Suppose next time, the man brings his linear measure to the market and gets a pair of shoes with the measure's length. But when he is back at home and tries to put on his new shoes, he finds that they are too small for his feet. What shall he do? Remember that he believes in the linear measure rather than his own feet. As long as his shoes are of the right size, they are the right shoes, even they are not comfortable for his feet. If he cannot put his feet in the right shoes, then the only thing he can do is to cut a bit off his feet to fit the right shoes! We can see that the man's belief in his measure is not only inconvenient, it can also be harmful or even dangerous.

Population policy is a more complicated case, for we cannot try everything beforehand. But the spirit should be the same: we want a pair of comfortable shoes instead of a pair of shoes of right size. But the criticism from some Western scholars is this: comfortable or not, this pair of shoes is not right. We can see that there is a deeper disagreement behind these two ways of thinking: What kind of choices are ethical, good choices or right choices?

2. WHAT IS ETHICAL?

There are some very important concepts in Western ethics: rights, autonomy and principles. This is not only the tradition of Western ethics, but also the main feature of contemporary Western ethics. Locke, Kant and Rawls are the main figures in this tradition. Although the so-called "justice ethics" dominates contemporary Western ethics in a significant sense, it is not the only voice in Western society. Recently, Carol Gilligan argued from a feminist point of view that there can be a different approach to the ethical life than acting upon universal principles, and this different approach is certainly no less mature; in fact, Gilligan argues that this is a much better ethical approach (Gilligan, 1982). And H. Dreyfus and S. Dreyfus show us by their phenomenological approach that it is the intuitive response to the concrete situation which represents the highest stage of moral maturity (Dreyfus & Dreyfus, 1990).

This different voice plays a very important role in the debate about the ethical implications of Lawrence Kohlberg's cognitive model of moral development (Kohlberg, 1981, 1984). Kohlberg holds that the development of the capacity for moral judgment follows three stages. The first stage is the Preconventional Level, in which the agent tries to satisfy her needs and avoid punishment; the second stage is the Conventional Level, in which the agent conforms to stereotypical images of majority behavior; and the third stage is the Postconventional and Principles Level, in which the agent is guided by universal ethical principles. The third stage is considered the highest stage of moral development.

According to these stages of moral development, Kohlberg (1981, 1984) found that men are more mature than women in moral development. Gilligan argues that these findings are derived from a male-dominated point of view. In doing this, she examines a moral dilemma used in Kohlberg's studies. In the dilemma, it is asked if a man named Heinz should steal a drug which he cannot afford to buy, in order to save the life of his wife. Kohlberg found that morally mature men tended to say that Heinz should steal the drug because the right to life is more fundamental than the right to private property, while women seemed unable to deal with the dilemma in a mature way. For example, a girl named Amy tried to solve the dilemma through communication. She thinks that maybe Heinz could go to talk to the pharmacist and gain the pharmacist's empathy so that the pharmacist could give him the drug. It is true that we cannot see any clear principle in Amy's resolution. But Gilligan points out that Amy's moral judgment is not immature, but rather represents a different way of thinking. That is, Amy recognizes the relationship between people and believe that communication can be a solution to conflict. Gilligan also tries to help Amy to articulate her principle: "If someone has something that would keep somebody alive, then it is not right not to give it to them" (Gilligan, 1982). Although in some sense this undermines the different voice she advocates, as Dreyfus points out rightly (Dreyfus & Dreyfus, 1990), it shows us the significance of the new ethics: that ethical behavior should be active instead of passive. An individual should not only do no harm to others, but should also care about others.

We can see that Amy's principle is just to make our world a better one; her highest moral principle is to do the good thing. Moreover, moral maturity is not only the ability to solve moral dilemmas successfully, but also to have less moral dilemmas, and to be able to stay in a situation and keep one's ethical intuitions once one faces a dilemma.

More importantly, Gilligan found a new way of thinking in Amy's solution to the dilemma, which is fundamentally different from "principle" ethics. This is what she calls "the different voice", which is does not represent a certain sex. She means by this to distinguish two kinds of voices in ethical judgment: justice ethics and care ethics. According the first point of view, good things are those that accord with the fundamental principles, while the second point of view says that good things are good just because they are good; their goodness is not derived from any principle.

Gilligan's work is important at least in three aspects. First, moral maturity is not the ability to use general principles, because falling into an ethical dilemma itself implies that the individual is not mature enough. Secondly, there is no general principle to employ. No two situations or people are exactly the same. Even a single individual is constantly changing. So it is impossible to know how to respond to a particular situation from several general principles. We may have different responses to situations from the perspective of care ethics, while justice ethics might have only one response. Thus a *good* choice might seem *unjust*. Thirdly, care ethics is a more active ethics while justice ethics is more passive. Care ethics demands not only that an individual do no harm to others, but also requires that the individual care about others' interests. This certainly points to a much better society.

H. Dreyfus and S. Dreyfus distinguish the same second voice. They prove from a phenomenological approach that "The highest form of ethical comportment is …to consist in being able to stay involved and to refine one's intuitions" (Dreyfus & Dreyfus, 1990). In saying this, they are not "denying that the ability to ask what is right reveals a kind of maturity", they just "see no reason to claim it is the *telos* of ethical comportment" (Dreyfus & Dreyfus, 1990).

Dreyfus and Dreyfus start their work from the research on driving students and chess players. They found that "It seems that beginners make judgments using strict rules and features, but that with talent and a great deal of involved experience the beginner develops into an expert who sees intuitively what to do without applying rules and making judgments at all" (Dreyfus & Dreyfus, 1990). The most important thing for the Dreyfuses to point out is that

> If the skill model we have proposed is correct and if everyday ethical comportment is a form of expertise, we should expect ethical expertise to exhibit a developmental structure similar to that which we have described above. On analogy with chess and driving it would seem that the budding ethical expert would learn at least some of the ethics of her community by following strict rules, would then go on to apply contextualized maxims, and, in the highest stage, would leave rules and principles behind and develop more and more refined spontaneous ethical responses (Dreyfus & Dreyfus, 1990).

They have a good example to explain the process:

> To take a greatly oversimplified and dramatic example, a child at some point might learn the rule: never lie. Faced with the dilemma posed by Kant—an avowed killer asking the whereabouts of the child's friend—the child might tell the truth. After experiencing regret and guilt over the death of the friend, however, the child would move toward the realization that the rule, 'Never lie,' like the rule 'Shift at ten miles per hour,' needs to be contextualized, and would seek maxims to turn to in different typical situations. Such a maxim might be, 'Never lie except when someone might be seriously hurt by telling the truth.' Of course, this maxim too will, under some circumstances, lead to regret. Finally, with enough experience, the ethical expert would learn to tell the truth or lie, depending upon the situation, without appeal to rules and maxims (Dreyfus & Dreyfus, 1990).

The two Dreyfuses' work is very important in the sense that they view ethical behavior as a kind of skill or ability to reach a certain kind of aim, not as something which has abstract value. Thus a closely related import of their work is that they do not view this intuitive moral judgment as something totally distinct from principled judgment, but treat these two as different stages of the same developmental process, with the former higher than the latter. That is, principled moral judgment does represent a certain kind of maturity, but it is the contextualized intuitive moral judgment which does not appeal to principles which represents the highest level of moral maturity. This is the difference between the two Dreyfuses and Gilligan. The difference is of great import, but will not be discussed in the context of this paper. I thus call the ethics proposed by Gilligan and the two Dreyfuses ("the second voice") the "good" ethics. This second voice holds that a mature moral judgment is not "acting upon rational judgments of rightness", but "intuitively doing what the culture deems good" (Dreyfus & Dreyfus, 1990). The first voice is thus called "right" ethics.

I now begin to analyze three differences between good ethics and right ethics.

2.1. Autonomy

A fundamental hypothesis of right ethics is that an individual is separate from others
and that she chooses the principles independently. On the contrary, the fundamental
hypothesis of good ethics is that an individual is immersed in the network of
interpersonal relationships and her moral concern is to retain these relationships.

We can see that when the boy suggests that Heinz should steal the drug to save
his wife, he views Heinz' behavior as something isolated, which has nothing to do
with other people. He ignores not only the direct consequence that the owner of the
pharmacy would lose his property, but also the indirect symbolic effect that if this
choice is right, then it would imply that stealing is tolerable or even ethical, if only
you think that you have sufficient reason to do this. We know that what stealing can
deprive someone of is not just property, but also the sense of safety and our belief in
society, and of course some people's belief in Heinz himself. So Heinz' behavior is
absolutely not isolated, but rather is deeply related to the people concerned and even
to the whole society. And he should not decide only between two abstract principles.
What he should do is consider the concrete situation, and he does not need to choose
between stealing and the death of his wife. It is absolutely possible for him to find
much better choices.

2.2. Right

Right ethics is a kind of principlism according to which our ethical judgment is
derived from some fundamental, universal ethical principles.

We can see easily that principles play a very important role in our ethical
judgment and everyday life. To see the import and limit of principles, let's consider
an everyday life example. Suppose we have

> *Principle A*: Do not jump into deep water if you cannot swim (or you
> endanger your life).

This is obviously a good principle. If you stay with this principle, you keep
yourself from danger. But now suppose you are in an exploding plane over the sea,
and the only way to save your life is to jump into the sea. What shall you do now?
Shall you obey principle A and stay in the plane? Of course not. You should
certainly put on your life jacket and jump into the sea as soon as possible. Then why
should you gave up that very good principle A? A simple answer is this: principle A
is now not applicable. But this does not imply that principles are not important in
our life generally. For we can say that in this circumstance we are using

> *Principle B*: Do not stay in an exploding plane (or you will die).

This is to say that although you give up principle A, you still have some other
principle to stay with. Then why principle B instead of principle A? The answer is
also very simple: Principle B gives you more opportunity to survive. That is, there is
another principle which is more fundamental than Principle A and Principle B:

> *Principle C*: Try to survive as long as possible.

We can see that Principle C is already very fundamental and abstract. It explains why we usually stay with Principle A and Principle B. If no reasonable choice violates this principle, we may believe that principlism is right. But now suppose that you are the person who is to explode the plane and that the only way to do this is to stay with the plane to the last minute. Thus the best choice for you is to violate principle C and stay with the exploding plane. Then here comes the principlist's problem: what makes you give up principle C? The reason is very simple: your aim is to explode the plane instead of survive, thus principle C is not applicable for you. Is there any other principle to explain your choice? The last hope of the principlist is the most fundamental principle:

Principle D: Do whatever serves your purpose best.

Any rational person should not violate this principle. In fact, this is just the definition of rational behavior. The pity is that the principle tells us nothing about what to do in a particular situation. The real general principle is actually a useless principle. It is not the kind of principle which principlists claim can direct our choice. So the last hope of principlists is a delusion.

Things are the same in ethical judgments. "Do not hurt others" is a good ethical principle, and parents usually teach it to their children. But we can not stay with it without variety. When we are physically hurt by someone, we may not only hurt them, but also hurt them badly. Why are we sometimes justified to hurt, and sometimes not? The answer is very simple. Hurting or not hurting is not the crucial point of the problem. The point is whether or not the choice is good for the agent concerned. In arguing against principlism, I am not saying that we should not use some ethical principles in our everyday life, nor am I denying that these principles are effective or that they play an important role in ethical analysis. What I am arguing is that no principle can be absolute, abstract or inviolable. The most important thing to consider is whether they are good for the agent concerned.

It might be of some help to consider several fundamental principles in justice ethics, such as beneficence, nonmaleficence, respect for persons and justice. Beneficence and nonmaleficence are the fundamental concerns I just mentioned, and we can put them together as the basis of further analysis. The principle of respect for persons may seem to be independent from the concern of goodness at first glance, but it is actually based on the latter. We respect persons because it is good for ourselves, for others and for the whole society, although the goodness is not always that obvious. The principle of respect for persons is not derived from a kind of tautological justification like "because they are persons, we must respect them," but is instead based on a certain kind of practical consideration. On the one hand, respect for persons can prevent one from doing something that is only good for herself; on the other hand, it prevents one from doing something just because she thinks it is good for someone else without considering the other person's willingness. The principle of justice requires that an ethical choice is good for everyone, not just for some people. So both the principle of respect for persons and the principle of justice are based on the principles of beneficence and nonmaleficence, which actually says that we should do good to the agent concerned. If there is someone who is trying to commit suicide, what we first think about is how to save her life instead of "respecting" her and leaving her alone to end her own life.

If some group of people is suffering, what we should do is to stop the suffering instead of having other people suffer to get justice.

2.3. Universalism

I think that one reason principlism is emphasized is that Western ethical criteria are universalized in a significant way. Western society already has a well-developed ethical network which is based on several ethical principles. Some Western philosophers thus think that these principles can be used to analyze all ethical problems, are applicable to all social situations, and are thus general, abstract, and inviolable. And in the view of some Western philosophers, we must insist on the universality of principles so as to analyze the ethical problems of developing countries correctly.

In analyzing ethical problems of developing countries, Western philosophers have found that many inhumane activities in those countries have very deep cultural reasons. Circumcision and the illegality of abortion, for example, are both products of certain cultures or religions. Some Western philosophers think that the only way to argue against these wrongdoings is to emphasize that the ethical principles of Western society are general, intercultural and inviolable. I believe this is the main motivation of principlism. But they have not realized that their opponents are just like them in the sense that they are also principlists. They think that those principles endorsed by their *own* culture should not be violated. Both of them think that principles should not be violated. The only difference is that they believe in different kinds of principle.

Without this realization, Western philosophers give very implausible arguments against inhumane actions such as circumcision. To put it roughly, the typical Western argument is this: our principles are general and absolutely right, but what you are doing conflicts with our principles, so you must be wrong. This kind of argument gives rise to the question: why are your principles are generally right while those principles endorsed by our traditional culture are wrong? In fact, Western philosophers can certainly admit frankly that even their principles were not always there; they have experienced a very complex process to reach today's recognition. From the illegality of abortion to its legitimization, from the merchandising of black slaves to their liberation and to the movement of equal treatment of all kinds of races, all of these have experienced and are going to experience a very long process. Thus a better argument for Western philosophers is this: principlism is wrong. No principle is or should be unchangeable. The wrong elements of principles of any culture should be abandoned as people pursue a better life; adhering to unreasonable principles would only make at least part of the people endure the misery of life. According to this argument, our principles are always changing and will continue to change, and of course your principles are also changeable; the change of ideas is the mark of social progress. And the justification for the change is its goodness for human society.

We can also see the contradiction in the Western argument. They first view those countries that have circumcision as some species extremely remote from us, and then go on to say to this remote species that we are right—not only for ourselves, but also for you. The contradiction is, given the fact that they are a remote species, how can what is right for us be right for them? In fact we first need to realize that although circumcision is only done in a few countries, it is only one form of the repression of women. We have circumcision in Africa, and we used to have foot-binding in China. We also used to have waist-binding in France. While African women are accepting circumcision, women in Western countries and other areas of the world are enduring every other kind of suffering, and they are usually considered to be willing. So the attack against circumcision should not be the attack of developed culture against undeveloped culture, it should be an attack against the whole culture which prejudices women, and this prejudiced culture exists deeply in all kinds of traditional culture.

3. CHINESE POPULATION POLICY

As to the problem of what is ethical, we can conclude with at least three points. First, ethical choice should be a good choice, not a choice derived from any principles. So the highest stage of moral judgment is to find the good choice intuitively instead of adhering to any kind of principles or making a difficult choice between several conflicting principles. Secondly, no individual is totally autonomous, separated from others. So an active ethics require that an individual not only do no harm to others, but also care about the interests of others. Thirdly, no ethical principle is generally applicable; ethical choices are those fit for concrete situations. With this in mind, we now turn to the population policy problem.

The first point just mentioned, that is, that ethical choice is just a good choice, has direct relevance to what I said in the beginning about the Western criticism of Chinese population policy (that this birth control policy is a violation of reproductive rights and is thus not ethical). I do not want to say here that because right ethics is not the best ethics, the criticisms derived from it are not effective. What I want to say is that, even if right ethics is unproblematic, the criticism of birth control policy is still problematic. Right ethics can not raise forceful criticisms against birth control policy. According to right ethics, some human rights can be more fundamental than others. For example, the right to survive can be more fundamental than the right of reproduction. Just like we can think that Heinz should steal the drug, we can also think that it is ethical to violate the right of reproduction, as long as we have enough reason to think that the right of reproduction conflicts with some more fundamental rights, say, the right to survive (Gilligan, 1982). Moreover, the critique is not only ineffective, but also misleading. From the perspective of right ethics, we can not see the necessity of education in the carrying out of population policy, and it is difficult to realize the importance of necessary medical service. And thus the population policy problem becomes a problem of choosing: once you have got the right choice, everything is fine. In fact, ethical behavior is not only the problem of choosing, but also a problem of how to carry out the choice. No matter how right the choice is, if it is difficult or even impossible to carry out in the current circumstance, then it can not be deemed a good choice. Even if we admit that Heinz is justified in stealing the drug, if he would have to kill the owner of the pharmacy or risk his own life in order to steal the drug, then he should not choose to steal the drug. And ethical judgment does not stop when we have made a choice, it is rather at the very beginning of its work. We not only need to judge *whether* we should carry out birth control policy, but also need to judge *how* to carry out the population policy, and how to make population policy best contribute to our life quality.

Secondly, some people think that the right of reproduction is a kind of fundamental right and should not be violated because they think that reproductive behavior is one's own business, affecting no one else, and thus should not be interfered with. But in fact, in a country with a very large population, one's reproductive behavior can have a very significant effect on others. Having more children not only make the population bigger, but it also has the important symbolic effect of implying that there is nothing wrong with having more children if you want to and are able to. We can see that if an individual does not consider these elements when she makes reproductive choices, her choice can not be judged as ethical.

Thirdly, one fundamental error of right ethics is that it holds that those ethical principles which are taken for granted in one circumstance should also be applicable in other circumstances. As to the population policy, because the Western population is not that big, the right of reproduction is taken for granted, and it is difficult to understand why the right should be limited. But in fact, no right is really endorsed by God, no right has existed from the very beginning of human history; rights can only exist under certain circumstances. An example might help to explain this.

When a child from a rich family hears that a poor child only has several pieces of bread a day, she might be surprised and even angry. She might tell the poor child that it is unfair, that to have enough food is one's fundamental right, which should not be violated. The poor child might agree with her immediately, and go to his family and tell them the same thing. His family might be excited with this: they are born to have enough food, they have the right! So they begin to eat as much as they want. But soon the ethical life turns out to be problematic: they eat all of their food for one month in just half a month. They will have nothing to eat in the second half of the month, and they may die from starvation! To enjoy the right of having enough food, they end up dead (Tang, 1996).

What is the problem? Is having enough food not the right thing? Of course not. The problem is that in a poor family, an ethical or rational choice is not to eat that much, if we are not determined to fight against the right to private property. We can see from this that our understanding of right is not universally applicable.

It is the same thing for reproductive rights. An obvious fact is that the Chinese population condition is totally different from that of Western countries, so the Western understanding of reproductive rights might not applicable to the Chinese situation. If we want to find a population policy which is applicable to the Chinese situation, we must study the current Chinese situation carefully and not let Western principles of "right" stop us from making good choices.

Chinese Academy of Social Sciences
Beijing, China

REFERENCES

Chinese classics:

Han Fei Zi: Wai Chu Shuo Zuo Shang.
Huai Nan Zi: Shuo Lin Xun.

Dreyfus, H. & Dreyfus, S. (1990). 'What is morality? A phenomenological account of the development of ethical expertise.' In: D. Rasmussen (Ed.), *Universalism vs. Communitarianism* (pp. 237-264). Cambridge: MIT Press.

Gilligan, C. (1982). *In a Different Voice: Psychological Theory and Women's Development*. Cambridge: Harvard University Press.

Habermas, J. (1982). 'A reply to my critics.' In: *Habermas Critical debates* (pp. 253, 244). Cambridge: MIT Press.

Kohlberg, L. (1981). *The Philosophy of Moral Development: Essays on Moral Development*, 1. San Francisco, California: Harper & Row.

Kohlberg, L. (1984). *The Psychology of Moral Development: Essays on Moral Development*, 1. San Francisco, California: Harper & Row.

Tang, R.-F. (1996). 'Reproduction, survival and development.' In: R-Z Qiu (Ed.), *Reproductive Health and Ethics* (in Chinese) (pp. 227-230). Beijing: BMU & PUMC Joint Press.

CHAPTER 17

REN-ZONG QIU

DOES EUGENICS EXIST IN CHINA?
ETHICAL ISSUES IN CHINA'S LAW ON MATERNAL
AND INFANT CARE*

This paper provides some background information and my view on the ethical issues in drafting and implementing the Law of the People's Republic of China on Maternal and Infant Health Care (abbreviated as "the Law" below).

1. IS THE LAW A EUGENIC LAW?

The Law has received much attention and criticism in scientific publications and mass media in the West. Some of the criticism is to the point, and some is not. Part of the controversy can be traced to misunderstandings due to cultural and linguistic barriers.

The term "eugenics" has changed its commonly understood meaning since the time of Francis Galton, Darwin's cousin. Today, eugenics is recalled as a practice in Nazi Germany where medical genetics was misused to pursue a policy of ethnic cleansing and racism. It is also recalled as a governmental program in the USA and other Euro-American countries where compulsory sterilization was used to try to reduce the population with physical, mental or social abnormalities. In their "Guidelines on Ethical Issues in Medical Genetics and the Provision of Genetic Services," Dorothy Wertz and her colleagues argued that "prevention is not eugenics":

> Today the word eugenics usually has a negative connotation, aligned with genocide. Most professionals reject the term outright in the context of medical genetics. To most people, eugenics means a social program imposed by the state. This is an approach to which people around the world object, because it denies human freedom, devalues some human beings, and falsely elevates the reproductive status of others.
> ...
> Medical genetics has as its goal the good of individuals and families. The ethos in present day medical genetics is to help people make whatever voluntary decisions are best for them in the light of their own reproductive goals. This is the decisive difference between present day medical genetics and yesterday's eugenics (Wertz, 1995, p. 3).

R.-Z. Qiu (Ed.), Bioethics: Asian Perspectives, 185-196.
© 2003 *Kluwer Academic Publishers. Printed in the Netherlands.*

The core of their argument is that eugenics is a program imposed by the state, regardless of the preferences of those individuals concerned, whereas prevention of genetic diseases is a voluntary choice made by particular individuals. For Wertz and her colleagues, informed consent is a criterion with which eugenics can be distinguished from prevention. In my opinion, a health program can be eugenic even if it requires informed consent. What makes a program "eugenic" in the sense of the term which we associate with the Nazis is racism. But I would like to use the term "eugenics" in the way Wertz and her colleagues used it for the convenience of discussing our topic in this paper, though the usage of the term 'eugenics' is still controversial.

Is the Law a eugenic law? Let us examine its structure. There are 7 chapters and 39 articles in the Law. 4 of these articles may have caused it to be viewed as eugenic. The others deal with the responsibility of the government at various levels to provide pre-marital and perinatal care (19 articles), and with procedures and qualifications (7 articles), incentive (1 article), ethical requirements (3 articles), legal liability (3 articles), definition (1 article), and time of implementation (1 article) (*Law of the People's Republic of China on Maternal and Infant Health Care*). I shall discuss three groups of articles:

(1) Ethical requirements: The Law's ethical requirements are in accordance with conventional ethical principles. Article 19 requires consent or proxy consent for any termination of pregnancy or application of ligation operation,[1] Article 32 prohibits sex identification of a fetus (for non-medical reasons), and Article 34 preserves keeping confidentiality.

(2) Legal liability: The Law prescribes punishment for standard offenses: practicing without proper license and qualifications; harming clients; and fraud. Only practitioners, not their clients, are subject to punishment. This provision is clearly stated in the authoritative 'Answers to Questions about the Law on Maternal and Infant Health Care' (p. 330).

(3) We come, finally, to the provisions which might suggest eugenic intent. Article 18 prescribes that physicians shall advise married people to terminate a pregnancy if a disease or defect is detected through prenatal diagnosis. But what the physician is required to offer is advice only. The decision to terminate pregnancy, as Article 19 prescribes, must still be made by the couple (or, if they are incompetent, their guardian). It should be kept in mind that Chinese generally view the fetus from a Confucian perspective: not yet a human being. "Birth is the beginning of a human being, and death is the end of a human being" (*Xun Zi*). Article 9 is similar: it directs physicians to provide medical advice to the couple, but what they provide is still advice that may be beneficial to the couple as case 3 below shows.

Article 10 is ambiguous. The official English version[2] says:

> After pre-marital medical examination, physicians shall, in respect of the male or female who has been diagnosed with certain genetic disease of a serious nature which is considered to be inappropriate for child-bearing from a medical point of view, explain the situations and give medical advice to both the male and the female; those who, with the consent of both the male and the female, after taking long-term contraceptive measures or performance of ligation operations, are unable to bear children, may get married. However, the circumstances under which marriage may not be contracted under the Marriage Law of the People's Republic of China shall be excepted (LPRCMIHC, p. 18).

This article raised four major issues for question and debate:

(1) What does it mean by "genetic diseases of a serious nature"? In Article 38 of this Law it is defined as follows: "Genetic diseases of a serious nature" refer to diseases that are caused by genetic factors congenitally, that may totally or partially deprive the victim of the ability to live independently, that are highly possible to recur in generations to come, and that are medically considered inappropriate for reproduction (LPRCMIHC, p. 28). This is the meaning of "genetic diseases of a serious nature", but not its reference. People (including geneticists) still do not clearly know which specific diseases do belong to this category. From a scientific point of view, in the majority of cases geneticists at most make a probabilistic prediction of a woman giving birth to a child with a certain genetic disease, but they do not exactly know what will actually happen.

(2) Does it mean that the medical advice which physicians provide becomes a legal directive? If so, physicians who do pre-marital examinations would take too much power in their hands. On the other hand, they would also take too much legal accountability: if the "medical advice" turns put to be wrong, the clients may sue her/him in court, as Case 1 shows:

> A clinical geneticist S in U Hospital in Beijing made a prenatal diagnosis for a woman client who was pregnant with twins. One member of the client's family is dead. After prenatal diagnosis, S told her that she "could deliver". After delivery the twins were found both dead. The client sued S to the court and requested half million Chinese dollars (CNY) in compensation. The case is still in trial, though the possibility of winning the case is slim.

There are a number of cases of this sort that are in trial. It is predicted that if "medical advice" or genetic counseling becomes directive, there will be much more cases of this sort.

(3) If Article 10 is interpreted as above, then it will be inconsistent with Article 19. Article 19 implies the principle of informed consent, but if medical advice becomes a legal directive in Article 10, it will not require observing this important principle.

(4) How should cases of noncompliance be dealt with? Suppose that a couple still gets married without taking long-term contraceptive measures or performance of ligation operations after medical advice was provided by the physician. Because there is no punishment of the couple, it will make the Article unpractical.

Article 16 is ambiguous, too: "If a physician detects or suspects that a married couple in their child-bearing age suffer from genetic disease of a serious nature, the physician shall give them medical advice, according to which the said couple shall take corresponding measures" (LPRCMIHC, 20). What are "corresponding measures?" Does the Law require agreement with what the physician advises? The Chinese text does not require strict compliance, though parents do have to take the physician's views into account in reaching their decision.

These two articles may cause a Westerner to suspect that the Law is eugenic in nature. However, I do not think it is fair to call the Law "eugenic" in its intent on the basis of one or two articles which raise suspicions. Of course, I do not mean that there are no problems in the Law. I will discuss these later.

The Chinese word *yousheng* has dual meanings. It may refer to the eugenics of the kind practiced by the Nazis during the Second World War, but it can also be understood to refer to a healthy birth. In today's Chinese language, "*yousheng*" routinely means "healthy birth", and is always used with "*youyu*", which means "rearing healthy children". In Chinese documents and media it is more often used to refer to healthy birth. The translation of Chinese "*yousheng*" into the English word "eugenics" caused misunderstanding and confusion. There is no racist tradition in China. Indeed, the Chinese were victims of racist slaughter by Western imperialists and Japanese militarists from the middle of 19th century to the middle of this century. Chinese generally do not think they are superior to any of the Western peoples or the Japanese, or that Han people are superior to ethnic minorities in China. The Chinese have made many grave mistakes over time, but racism has not been among them.

The differences between Chinese *yousheng* and Nazi Germany's eugenics, in my opinion, are these: (1) Racism was essential to Nazi Germany's eugenics, but has nothing to do with Chinese *yousheng*, which can be understood as the hope to reduce defective births. When the Chinese speak of the "quality of the population," there is no implied reference to any race or ethnic group.[3] In Chinese, "population" is represented as "the number of human mouths", i.e., the sum of individuals. The Chinese nation, moreover, consists of more than 56 ethnic groups. (2) Nazi eugenics was imposed without any informed consent, but in the Law informed consent is a principle which is to be observed in some Articles. The language used in some Articles is not fully explicit on this point, and should be improved.

2. MOTIVATIONAL FACTORS IN DRAFTING THE LAW

Several factors motivated the draft of the Law:

Officials: The Chinese revolution originated in remote, mountainous areas which were called Revolutionary Old Bases. Officials who returned to their villages where they had stayed during the war against Japanese occupation or civil war were shocked by their extreme underdevelopment. They were especially appalled by so-called "idiot villages" where almost all villagers were physically deformed or mentally retarded, none of whom were capable to serve as treasurer or as head of the village, and all of whom had to be completely dependent upon support from the community and government. In the Dabie Mountains, one of these Revolutionary Old Bases, the source of 37.5% of severe mental retardation was inbreeding. Many suffered from cretinism.

Unfortunately, many officials cannot distinguish between congenital diseases and genetic diseases. When I asked one of the officials in Gansu Province how one knows when severe mental retardation is caused by genetic factors, he replied that retardation is genetic if it persists through three generations. Officials there mistook the high prevalence of physical deformity and mental retardation as a cause of community underdevelopment, rather than as a consequence of it.

Geneticists: Chinese geneticists have played an important role in drafting the law. Together with some officials, their genuine concern was for the welfare of people, and they worried that some areas, indeed the country as a whole, are still trapped by poverty. They also wanted to use the law to promote genetics and its application in medicine. Their emphasis on genetic factors in determining human traits such as intelligence and behavior can be understood in part as a reaction to the biological doctrine of Lysenko, once favored by the Communist Party.[4]

At the same time, it must be said that geneticists have been insensitive to ethical issues. They confused what we can do technologically with what we ought to do ethically. During the workshop (Minutes, 1991), I argued with those who insisted that for the social good it should be compulsory for the severely mentally retarded to be sterilized. I said if I were a representative of the mentally retarded I would ask all of them, for the sake of the social good, to be compelled to donate CNY 50 per month to the community for improving their lot. They strongly protested this proposal, insisting that "you should obtain our consent." I asked why the donation of a bit of money for the social good requires consent from those affected, while an invasive operation on my body might be performed without the consent of me or my guardian.

Handicapped: Because of China's underdeveloped economy and its cultural traditions, there is little familial and social support for defective infants. In traditional China, a defective newborn was called a "monster fetus" (*guai tai*) and did not have the status of a human being. A woman who gave birth to a defective newborn was usually under financial, cultural, and familial pressure. These traditional Chinese attitudes toward deformed children are similar to those of the Japanese. Changing attitudes toward the defective child and its mother will take time. In the near future there is no hope to dramatically improve such an unfavorable societal and familial context. All of these make the life of a defective child and its mother miserable. Given the unchanged social and cultural context, the prevention of a deformed baby will benefit it and its mother.[5]

During May-October 1988 a national epidemiological survey on mental retardation of children aged 0-14 was held all over the country. The result showed that among 81,396 children, the prevalence of mental retardation was 1.07% (0.75% in urban areas, 1.46% in rural areas) (see Minutes, 1991). In 1989, the Chinese Association for the Handicapped, in a letter suggesting prompt adoption of measures to ensure healthy birth and prevent the birth of handicapped infants, revealed that the population of handicapped in China amounts to 51.64 million, 4.9% of the whole population. Of these, 10 million were congenitally handicapped, and among the 10.17 million mentally retarded, 50% were caused by congenital factors. In the letter it was claimed that

> deformity brings about sufferings to handicapped, misery to the family, and a heavy burden to the country and society. In assisting the government and mobilizing the society, China's Association for the Handicapped devotes itself to the cause of handicapped, calls the whole society to act in the spirit of humanitarianism, and to understand, respect, care and help handicapped, making them genuinely equal members in the society who enjoy rights of full participation.

From the same principle of humanitarianism, they insist that births of deformed children should be reduced.

> We call on the government and society to further strengthen the prevention of deformity, beginning with the prevention of congenital deformity and a reduction of the birth of miserable children. We maintain that the principle of humanitarianism is consistent with the improvement of the quality of our population ('A Letter to Suggest a Speed up for Healthy Birth Legislation for Preventing Congenital Disability', Second Session of the First Presidium, China's Association for Handicapped, 1989).

Lay persons: There are cultural, economic and policy reasons for lay persons to pursue healthy births. Pursuit of a healthy birth as well as a peaceful death has its roots in Chinese cultural tradition. The greatest Confucian (apart from Confucius and Mencius), Xun Kuang, said: "Birth is the beginning of a human being, and death is the end of a human being. A human being who has a good beginning and a good end fulfills the *tao* of mankind" (*Xun Zi*). In a moral sense, a good beginning and a good end mean that a human being passes her or his whole life in a moral way, and in a physical sense had best have a healthy birth and peaceful death. A survey made by my colleagues in the late 1980s found that if a couple has a deformed baby, one of them has to resign from work, and the extra cost will be one third or one quarter of a worker's average salary. Population control policies may be a factor too, though a couple with a defective child is permitted to have another child. Many geneticists and doctors are against this policy for they worry that it will increase the percentage of handicapped in the whole population.

With increased publicity of medical and genetic information, many young couples are anxious to seek prenatal diagnosis and genetic counseling. Even the future mother-in-law in the village knows of prenatal diagnosis and may accompany her future daughter-in-law to the clinic (though their chief concern may be less with genetic disease than with whether the hymen of her future daughter-in-law is intact). Demand for genetic services is increasing (Qiu, 1996).

3. TWO APPROACHES IN DRAFTING THE LAW

There have been two approaches in drafting the Law.

3.1. *"Eugenic"* Approach

One approach is to focus on reducing the population with physical deformities or mental retardation by limiting marriage and reproduction by compulsory means. An example is Gansu's Law, called the "Regulation on Prohibiting Reproduction of the Dull-witted, Idiots or Blockheads," promulgated by the Fifth Session of the Standing Committee of People's Congress of Gansu Province on November 23, 1988 (Lei *et al.*, 1991).

The definition of "dull-witted, idiots or blockheads" is:
1. Congenitally caused by familial inheritance, inbreeding or parents under external influence;
2. Mental retardation at middle or severe degree with IQ below 49;
3. Behavioral disorders in language, memory, orientation, thinking etc.

It prescribes prohibiting the reproduction of dull-witted, idiots or blockheads. The dull-witted, idiots or blockheads may be married only after sterilization. If a couple are both dull-witted, idiots or blockheads, only one has to be sterilized; if only one of the couple is dull-witted. idiot or blockhead, only he or she must be sterilized (Lei *et al.*, 1991, pp. 131-132).

Gansu's Law is indeed a eugenic law, similar to those enacted in the USA or other Euro-American countries before the Second World War. Unfortunately, the contributions of some colleagues in medical ethics and health law in developing the Law account for some of its objectionable features. First, they labeled physical deformity and mental retardation "inferior birth," a term which is prejudicial toward the handicapped. Some even claimed that medical care promotes inferior birth and impedes the struggle for survival (He *et al.*, 1989; Wu *et al.*, 1990; Zheng, 1990; Chen, 1992). This term was also used in some governmental documents and ministers' speeches. I argued against the use of this term in (Qiu, 1992). Secondly, they referred to those parents with severe genetic diseases, severe schizophrenia, inbreeding and advanced age as "parents without reproductive value" (He, Z-X, 1985; He, L *et al.*, 1989; Wu *et al.*, 1990), which is also a prejudicial term. Thirdly, they favored an ideology in which the priority was put on social good or state interest regardless of individual interests (Du, 1985; Zhu, 1987; Chen, 1992; Zhan, 1996). Their thinking can accurately be characterized as eugenic.

3.2. *"Healthy birth" Approach*

The other approach is to focus on improving pre-marital and perinatal health care with the purpose of reducing misery and enhancing the happiness of the family and individuals. It provides information, education, counseling and services to prevent severe congenital or genetic diseases and respects the decision made by family and individuals. The reduction of handicapped in the population is only an indirect consequence of these individual and familial decisions. Informed consent is a crucial principle in this approach apart from other fundamental bioethical principles. These goals and principles are endorsed in the Minutes of the First National Workshop on Ethical and Legal Issues in Limiting Procreation. (Minutes, 1991) The main points of the Minutes are:

Medico-genetic perspective. In the etiology of mental retardation, 82.3% is made up by non-genetic congenital or environment factors. Participants agree that the sterilization of those mentally retarded people whose condition is caused by genetic factors may reduce the incidence of mental retardation by a small amount, but the focus should be on developing pre-conception and perinatal care, maternal and child care in a program of community development with the goal of effective prevention of mental retardation.

Ethical perspective. Participants agree that any proposed action which would limit procreation by the severely mentally retarded must be evaluated on the basis of ethical principles of beneficence, respect, justice and solidarity. In some cases, sterilization of the severely mentally retarded may be in their best interest. They should enjoy the same rights as others. Their family, community and society ought not to discriminate against them, and ought to protect their rights rather than infringe upon them. The workshop calls for the whole society to identify with physically and mentally handicapped compatriots, do the best to provide social support for them and respect their right to life. Limiting their ability to procreate can be ethically justified only when procreation would be harmful to them. Because the severely mentally retarded are incompetent, they have no responsibility for the various consequences of their mental retardation; accordingly, limits on their procreation, such as sterilization, must not be any kind of punishment, but rather a contraceptive measure to reduce the misfortune of them and their families. Participants agree that because of their incompetence, the severely mentally retarded are unable to make rational judgments on what is in their best interests, and must rely on the decision of a guardian or proxy who has no financial or emotional conflict of interest.

Legal perspective. A law of compulsory sterilization would infringe upon civil rights laid down in the Constitution and other laws, including personal rights and an incompetent person's rights under guardianship. A law providing for voluntary sterilization would respect these rights (Minutes, 1991).

In (Chen *et al.*, 1989) the author emphasized that women have the right to decide on the termination of pregnancy or sterilization (pp. 261-262) and in (Chen *et al.*,

1992) the author claimed that it is necessary to obtain consent to sterilization and termination of pregnancy (p. 205). In (Feng, 1994) the author pointed out that the proponents of eugenics tend to ignore the difference between genotype and phenotype, and give the priority to the nation over individuals (pp. 246-247). She also argued that it is ethically justified to sterilize the severely mentally retarded only if it is in their best interests, that it should be voluntary but not compulsory, and that consent or proxy consent should be obtained both to sterilize the severely mentally retarded and to abort a fetus with a genetic defect (pp. 250-253).

4. EVALUATION OF THE LAW

The Law is a result of compromises between these two approaches. As a whole the second approach prevails over the first one in the sense that the focus is put on the commitments of the government to provide pre-marital and perinatal health care to ordinary people, specifies the right to informed consent and confidentiality, and prohibits sex selection for non-medical reasons with the use of prenatal diagnosis. Of course, there are still problems that require attention.

In my opinion the principle of informed consent should be listed in the first chapter of "General Provision." This would issue an explicit and unambiguous message to all people concerned that the purpose of the Law is to provide information, counseling, techniques and services to clients and help them to make decision on marriage and reproduction.

In Article 9 it is mentioned that those who are infected with any target disease shall postpone their marriage for the time being. In Article 38 it is specified that this provision refers to AIDS, gonorrhea, syphilis and leprosy (LPRCMIHC, p. 27). HIV, however, is a lifelong infection and remains infectious. Thus Articles 9 and 38 bar marriage to HIV-positive individuals. These articles have to be changed.

Article 10 should be revised. If what physicians provide is only advice or medical advice, it should be explicitly specified that the decision of whether to marry should be made by the clients, upon learning that one of them has a certain genetic disease. And if Article 10 is interpreted as prohibiting the marriage of clients one of whom has a certain genetic disease, it will be in conflict with China's Constitution and other laws. The Constitution stipulates that marriage, family, mother and child be protected by the state. China's General Principles of Civil Law provide for legal protection of the rights of the handicapped, and China's Law on Safeguarding the Handicapped stipulates that the handicapped enjoy the same political, economic, cultural, social and familial rights as other citizens, that the civil rights and personal dignity of the handicapped be protected by law, and prohibits discrimination, stigmatization, bullying or humiliation of the handicapped (*Collection of Laws*, pp. 1-21, 76-82, 188-205). Moreover, In "Answers to Questions about the Law on Maternal and Infant Health Care" (p. 331), it is explicitly stated that there is no compulsory regulation on marriage and reproduction in the Law. If so, the wording of Article 10 has to be changed.

As pointed out above, the definition of "genetic diseases of a serious nature" in Article 10 is inadequate (LPRCMIHCL, p. 28). Geneticists cannot specify such conditions with certainty, nor can medical advice on marriage be formulated with sufficient certainty. Hence the wording of Article 10 should be changed.

In Article 16 it should be explicitly clarified that any physician's medical advice is non-directive.[6]

5. ETHICAL ISSUES IN DRAFTING AND IMPLEMENTING THE LAW

Any action taken in the draft and implementation of the Law should be evaluated in an ethical framework which consists of basic principles of bioethics: nonmaleficence and beneficence, respect, justice and solidarity.

5.1. Welfare of Individuals and Family vs. Reduction of Handicapped Population

I will not elaborate on this issue, having discussed it above. Here I add merely that the purpose of the Law is to improve human existence through the prevention of genetic diseases. It is for the welfare of individuals and families; reduction of the handicapped population is an indirect consequence of the prevention.

Case 2:

> In a pre-marital examination Nana was told by the physician that she suffered from an ovarian cyst and she should receive regular check-ups to avoid its becoming malignant. After marriage she went to her clinic to be checked up regularly and was treated for two years. She then gave birth to a baby by cesarean section (Zhan, 1996, p. 31).

Case 3:

> L and his fiancee S planed to be married. After a pre-marital examination L was found to suffer from an undisclosed disease. If they have sex, the disease will make them infertile. The couple postponed their marriage. After 8 months of treatment, L's disease was cured. They then married and gave birth to a child one year thereafter (Zhan, 1996, p. 31).

Many officials do not understand the patterns and mechanisms of human heredity, including the phenomenon of natural mutation, and they greatly exaggerate the power of controls on fertility in reducing the incidence of mental retardation. But the physicians who attended a training workshop and who would conduct pre-marital examinations for the couples who get married did not lack this knowledge. Before my talk about the ethical issues in the Law, I asked them two questions: (1) If my parents are both healthy and all of my relatives are healthy, can I conclude that I have no disease caused genes? (2) If all people with genetic disease decide to have no children, then will nobody suffer from genetic disease in the world? All physicians who attended my lecture answered "no": for (1), since I may be a carrier; and for (2), due to the additional reason of natural mutation.

5.2. Voluntary vs. Compulsory, Informed Consent vs. Imposition by the State

Everyone has a right of self-determination in marriage and reproduction, though there are minimal exceptions, such as inbreeding, incest, rape, and a couple of diseases. These are principles accepted by the international community. Under China's unique circumstances of overpopulation, the right to reproduction could be justifiably limited for a certain period. This circumstance, however, sets a limit beyond which any compulsory interference with reproductive choice cannot be justified. The purpose of the Law is for the welfare of individuals and their families, and accordingly any decision on marriage or reproduction should be made by the client herself or himself. Similarly, any medical counseling on postponing or withdrawing a marriage application, or on sterilization or termination of pregnancy, should be advisory only. The client's informed consent is mandatory, and no decision should be imposed by the physician or by the state.

There is no way to guarantee that the Law will not be used by somebody for eugenic purposes, but as I know, physicians are generally familiar with the principle of informed consent. I was invited to give a lecture on Ethical Issues in Marriage and Reproduction at a workshop to train those physicians who will be involved in pre-marital examination by the Bureau for Maternal and Child Care, Ministry of Public Health on September 21, 1996. My impression was that all 120 trainees who attended the workshop were of good quality medically and ethically.

5.3. Medical Advice: Facts vs. Values

The phrase "medical advice" occurs in the Law many times. But medical advice is a mixture of medical fact and also values, reflecting the physician's beliefs and attitudes. The physician's values, however, may not be identical to the clients'. Suppose, for example, that a millionaire is carrying a fetus with Down's syndrome. The physician may suggest that she terminate the pregnancy, reasoning that the disability of the child will become a heavy burden to the couple. The millionaire, on the other hand, may reply that she and her husband will love and care for the baby after delivery and that the cost of care will be not a financial burden. What would be the physician's reply, and what power ought to be given to the physician in such a case? Facts and values, and physician's values and client's values must not be confused with each other.

Case 4:

> A musician Y and an engineer W went to a clinic to undergo pre-marital examination. The physician told them that they have incompatible blood types, and that the incidence of newborn hemolysis is 3%. He suggested that they consider the possibility and consult their parents. They said they decided to marry. But just before wedding Y changed her mind: "After deliberation, I have decided not to marry W. Out of concern for his family, I want to avoid a marriage without offspring because he is a single son of his parents." The author of this report remarked that she is wise, because she believes in science and controls her passion by reason, preventing the possibility of a newborn's hemolysis (Zhan, 1996, p. 30).

But there are many options facing Y. The choice she made is only one among them. Which option she chooses should depend on her priorities and values.

Case 5:

> L and Y want to marry. After a pre-marital examination, L was found to be infertile. Y, upon learning this, wanted to break off with her. But she did not believe the diagnosis because she had always been healthy. She insisted on marrying Y, and Y did not refuse. After marriage she was proven again to be infertile. Y and his mother were anxious to have a baby. Their marriage was unhappy, and 10 years later they divorced. The author of the case report made a comment to the effect that if they had believed in science, their sad story would have not happened (Zhan, 1996, p. 30-31).

However, this is not a report about science, but about value. If Y had not given higher priority to having a child than to sustaining their mutual love, the story would have ended happily.

5.4. Confidentiality and Privacy

In implementing the Law, keeping confidentiality and privacy for clients is very important.

Case 6:

> A girl L and a boy X wanted to marry. After a pre-marital examination, the physician wrote "hymen was broken, had sexual behavior" in the form, and this was seen by X. X then refused to marry L. The author of the report made no mention of the need for the physician to maintain confidentiality and to respect the girl's privacy. Instead, he remarked that a girl would be happy to marry such a man, one who gave higher priority to a woman's hymen than to the woman herself (Zhan, 1996, p. 30).

Authorities now have a regulation prohibiting mention of the condition of a woman's hymen in the pre-marital form or on any other occasion.

China's recent *yousheng* endeavor involves two interventions which have to be mentioned here. One is to popularize iodine salt in the areas with high prevalence of cretinism. The other is to promote the ingestion of folic acid by fertile women in areas with high incidence of neural defects. These two initiatives show that the *yousheng* is now on the right track. I am quite optimistic about the future of the Law. Partly because of the criticism abroad, officials, geneticists, physicians and other personnel concerned will be more prudent. And they also have become more and more sensitive to ethical issues concerning the Law. At an appropriate time, the Law can be revised in the legislature. Western colleagues who are concerned about the Law should go to China and have a dialogue with Chinese officials, geneticists, physicians and laypersons and help them to do better. Protests, charges, threats or sanctions, in my opinion, will have no effect at all.

Chinese Academy of Social Sciences
Beijing, China

NOTES

* This article was presented at a meeting of the HUGO Ethical Committee in San Francisco, USA on November 25, 1996 and revised after the meeting. Thanks to Professor Hiraku Takebe and Professor Daniel Wikler for their careful reading of the manuscript and very good advice for revision. Of course, it is my responsibility for anything presented in this article.

[1] Article 19: Termination of gestation or performance of ligation operations practiced in accordance with the provisions of this Law shall be subject to the consent and signing of the person *per se*. If the person *per se* has no capacity for civil conduct, it shall be subject to the consent and signing of the guardian of the person.

[2] The English version of the Law circulated in the West is not very accurate. The official English version which I cite in this paper was authorized by the Law Committee of National People's Congress, China's legislature.

[3] So-called "one couple one child" is somehow misleading. Actually China's population policy is "one, two, three, four and no limit". For Han people (more than 90% Chinese are Han) in urban areas a couple should only have one child; for Han people in rural areas two children; for minorities three or four

children; and for Tibetans there is no limit. The policy is based on the population of a certain ethnic group. In 1949 the Tibetan population was only 1 million, but now it reached 2 million.
[4] Fairly speaking, the attitude of the Chinese Communist Party towards Lysenko was different from that of the Soviet Union. In 1956 at a conference on genetics the favorite attitude towards Lysenko was overturned. Please see (Li, P-S et al., 1996).
[5] Of course, it fails to avoid the paradox raised by Parfit (1984): Life which is not good, but is worth starting.
[6] In collaboration with my Chinese colleagues I recommended these opinions to the authority in different occasions. The responsible men of the Law Committee, National People's Congress and Ministry of Public Health invited two leading geneticists and bioethicists (I was one of them) to attend a meeting on August 4, 1998. At the meeting they informed us that our opinions were accepted by the authority, and the Law will be revised at an appropriate time.

REFERENCES

Chinese Classic:
Xun Zi

_____ (1995). 'Answers to questions about the Law on Maternal and Infant Health Care,' *Maternal and Child Care of Chin*a 10(6), 330-331.
_____ (1992). *Collection of Laws of People's Republic of China*. Beijing: Law Press.
_____ ()*Law of the People's Republic of China on Maternal and Infant Health Care* (LPRCMIHC). Beijing: Printed by the Ministry of Public Health of the People's Republic of China, Beijing.
_____ (1993). 'Minutes of the First National Workshop on Ethical and Legal Issues in Limiting Procreation, Beijing, November 11-14, 1991,' *Chinese Health Law*, 5, 44-46.
Chen, M-G *et al.* (Eds.) (1989). *An Introduction to Medical Law*. Shanghai Science & Technology Press.
Chen, M-G *et al* (Eds.) (1992). *Health Law*. Shanghai Medical University.
China's Association for the Handicapped (1989). 'A letter to suggest speed up the healthy birth legislature for preventing congenital disability,' the Second Session of the First Presidium, China's Association for the Handicapped.
Du, Z-Z (1985). *An Outline of Medical Ethics*. Jianxi People's Press.
Feng, J-M (1994). *Modern Medicine and Law*. Nanjing University Press.
He, L. *et al.* (Eds.) (1989). *Modern Medical Ethics*. Zhejiang Education Press.
He, Z-X (Ed.) (1985). *An Introduction to Medical Ethics*. Jiangsu Science & Technology Press.
Lei, Z-H *et al.* (eds.): 1991, *Handbook of Family Planning Administration*, printed by Gansu Commission on Family Planning, pp.131-132.
Li, P-S. *et al.* (1996). 'The Qingdao conference of 1956 on genetics: The historical background and fundamental experiences.' In: D. Fan & R. Cohen (Eds.), *Chinese Studies in the History and Philosophy of Science and Technology* (pp. 41-54). Dordrecht: Kluwer Academic Publishers.
Liu, B-R *et al.* (Eds.) (1987). *A Course on the Foundation of Law*. Shanghai Transportation University Press.
Liu, Benren *et al.* (Eds.) (1991). *A Course on the Foundation of Law Science*. Shanghai Medical University.
Parfit, Derek (1984). *Reasons and Persons*. Oxford: Clarendon Press.
Qiu, R-Z (1992). 'Ethical issues in the sterilization of severely mentally retarded,' *Chinese Medical Ethics*, 1, 10-15, 29.
Qiu, R-Z (1996). 'Genetic counseling in China," *Proceedings of the Third Session of International Bioethics Committee*, vol. II, UNESCO, 25-26.
Wertz, Dorothy *et al.* (1995). *Guidelines on Ethical Issues in Medical Genetics and the Provision of Genetic Services*, WHO, p. 3.
Wu, X-Z *et al.* (Eds.) (1990). *Modern Clinical Medical Ethics*. Tianjin People's Press.
Zhan, Z. (1996). 'The first step of sexual love,' *Happy Families*, 9, 30-31.
Zheng, R-S (1991). 'Healthy birth should be taken as moral concept and norm of human beings.' In: X-Z Wu *et al.* (Eds.), *Medical Morality: Theory and Practice* (pp. 200-205). Tianjin People' Press.
Zhu, X-F (1987). 'Moral consideration on eugenics, proceedings of the Fourth Conference on Medical Morality,' *Journal of Medicine and Philosophy*, 12, 43-47.

CHAPTER 18

YAN-GUANG WANG

AIDS, POLICY AND BIOETHICS: A NEW BIOETHICAL FRAMEWORK FOR CHINA'S HIV/AIDS PREVENTION

1. POSSIBILITY FOR CHINA TO BECOME A COUNTRY WITH A HIGH RATE OF HIV INFECTION

The first case of HIV infection was reported in China in June 1985. At the end of 2001, there were 30,736 reported cases of HIV-positive and AIDS patients all over the country. Experts estimated that the number of HIV-positive people has reached 600,000 (Beichuan Zhang, 2002, p. 1). Even though the numbers of HIV infection seem to be quite low considering China's population of 1.2 billion, we have reason to assume that it is very probable that China will become a country with a high HIV infection rate. The reasons are as follows:

Reforms in China's economic structure have given rise to population mobility on an unprecedented scale. The mobile population, which is estimated to reach 80-100 million, live far away from their homes, are without traditional cultural and moral constraints and can easily develop high-risk behavior, all of which makes it highly likely that they will be infected with HIV. They are traveling around the country with goods, services, information and diseases.

With economic reforms and social dislocations, a proliferation of all kinds of high-risk behavior has emerged, such as unprotected intravenous drug use, prostitution and homosexual activity, fostering the spread of HIV infection.

Other STDs which facilitate HIV infection have spread over the country rapidly. Premarital and extramarital sex, casual sex, and multiple sex partners have increased in the younger generation, again facilitating the spread of HIV infection. Moreover, there is a real danger of iatrogenic HIV infection, especially through blood transfusion.

Being faced with the danger of becoming a country with a high HIV infection rate, there are some misleading conceptions of HIV/AIDS which have run counter to the countermeasures that are necessary for effectively preventing and controlling the spread of HIV infection. For example, because the majority of Chinese do not know

R.-Z. Qiu (Ed.), Bioethics: Asian Perspectives, 197-205.
© *2003 Kluwer Academic Publishers. Printed in the Netherlands.*

how to protect themselves against HIV infection, some Chinese officials have claimed in the mass media that "AIDS is the plague of the 20th century" engendering public fear and panic. Some healthy educators and sexologists have even said that "AIDS is the punishment for promiscuity". As a result, HIV-positive and AIDS patients and also uninfected persons in high-risk groups suffer from discrimination and stigma.

Many Chinese officials and health workers believe that the conventional public health approach is sufficient to prevent or control the HIV epidemic. But HIV infection is an epidemic so special that the conventional public health measures, such as testing, reporting, contact tracing, and isolation, are not sufficient to deter the epidemic. Another important problem in China's HIV prevention is that the laws of prohibiting prostitution and drug use have made many prostitutes and intravenous drug users go underground, and they thus have no chance of receiving HIV education for changing their unsafe activities.

The problems in China's HIV prevention policy and law show that an adequate bioethical framework to evaluate the action taken by policy and law-makers on controlling the HIV epidemic must be formulated. It is indispensable for shaping an effective and supportive ethical and legal climate for HIV prevention in China and addressing the ethical issues.

2. FORMULATING A CURRENT BIOETHICAL FRAMEWORK TO DEAL WITH MORAL AND POLICY ISSUES IN HIV PREVENTION

Now which bioethical framework ought to govern China's HIV prevention policy? The basic bioethical framework we use to evaluate actions in bioethics consists of principles such as nonmaleficence, beneficence, respect for autonomy, and justice. It seems that these principles are not fully suitable for HIV/AIDS prevention.

I suggest an improved bioethical framework that consists of the principles of tolerance, beneficence, autonomy and care. In my suggested bioethical framework the principles of tolerance and care should play a central role.

The principle of tolerance is located in the first order of this bioethical framework. It generates other principles and facilitates dealing with the moral and policy issues in biomedicine, especially how to treat HIV/AIDS-related special population problems.

The basic bioethical framework, when applied in the ordinary medical context, is based on a presumption that all patients have equal social and moral status, but when it is used in the HIV epidemic, we encounter problems. In the HIV epidemic, some special social groups, such as AIDS patients, HIV positives, drug users, prostitutes and homosexuals are involved. They are marginalized and often stigmatized and discriminated against. It is very difficult for the public to presume that they have equal social and moral status. The public often thinks that AIDS patients and HIV-positives got the disease through immoral or illegal behavior, so they deserve punishment. The public often regards homosexuals, drug users and prostitutes as perverse, abnormal, impermissible or even disgusting, so their rights are often infringed upon. How can these persons, whom some feel ought to be punished, be treated by the principles of nonmaleficence, beneficence, respect for autonomy and justice? The basic bioethical framework ought to apply to them, but actually it has never fully been applied to them. As a result, the basic bioethical framework is inadequate for HIV/AIDS-related special social groups.

This problem is caused by the inadequacy of those principles as well as by the foundational theories of this basic bioethical framework. In the basic bioethical framework, the principle of justice seeks to provide the best possible health care for all citizens and promotes the ideal of equal access to health care for everyone, including care for indigents and the marginalized. However, it has never done this well.

The foundational theories of principles of justice are utilitarianism, libertarianism, communitarianism and egalitarianism. Utilitarian theories hold that justice is a form of obligation created by the principle of utility. Typically utilitarian obligations of justice are rights for individuals that should be enforced by law and which are contingent upon social arrangements that maximize net social utility. Apart from aggregate welfare, utilitarians neglect considerations of justice that focus on how benefits and burdens are distributed. In such a way, social utility might be maximized by not allowing health care for some of society's sickest and most vulnerable or marginalized populations.

Libertarian theories also have difficulty with justice for the marginalized. For example, Robert Nozick's "entitlement theory of justice" argues that a theory of justice should affirm individual rights rather than create "patterns" of economic distribution in which governments act to redistribute the wealth acquired by persons. Even though he thinks that the indigent will receive compensation from procedural justice, uncompromising commitment to liberty is dangerous to marginalized populations.

The inadequacy of foundational theories can cause very devastating effects for the marginalized, such as AIDS patients, HIV-positives, homosexuals, prostitutes, and drug users, whose status may be due to chance, accident, casual offense, or biological or social reasons.

In order to dispense justice to varied populations, especially in the HIV/AIDS-related population, and for the solidarity of all groups regarding HIV-prevention and social justice, some preconditions and new arguments might be added to the principle of justice in order to formulate a new and improved principle.

3. PRINCIPLE OF TOLERANCE

I suggest the principle of tolerance and justify it by some additional theories. The principle of tolerance means that members of a moral community should a) permit members of other communities to do what they themselves think wrong or do not want to do, b) permit members of other communities to have a lifestyle they themselves do not want, and c) forgive the mistakes of others. In almost every country, there are many moral communities who do not share the same beliefs or value system. Some populations, such as homosexuals, have a different lifestyle or set of moral views. The principle of tolerance is presumed on the basis that the differences or disagreements between different moral communities exist at all times, and that members of different moral communities have to live together as neighbors or co-workers and seek common ground while retaining their differences. The principle of tolerance will give members of different moral communities more freedom, build a more equal relationship among them and shape a more supportive network in society. For HIV/AIDS prevention, such a supportive network is very important.

To tolerate others, one must first respect them. The natural law can justify respect for others. It argues that human nature demands that humanity respect each other (Beauchamp, 1982, p. 308). The factor of human nature or the person's character which is different from other animals is self-consciousness, rationality and higher intelligence. The common nature of humanity makes humanity respect each other, just like the Chinese saying "All are brothers and sisters within the four seas."

Kant also justified respect for others. He argued that because humans are the only rational agents, they must be respected (Gillon, 1993, p. 16). Rawls's theory of justice explicates justice as fairness, understood as norms of cooperation agreed to by free and equal persons. He justifies his interpretation of justice by our deeper understanding of ourselves, our aspirations, and our realization of public life. His theory is based on this perspective, which seems to be commonly accepted by rational agents. He claimed that rational agents blinded behind a "veil of ignorance" regarding their personal situation would choose principles of justice that maximize the minimum level of primary goods in order to protect vital interests such as health in potentially damaging contexts. These agents would choose social allocations to meet certain needs. Social policy would guarantee a safety net or minimum floor below which citizens would not be allowed to fall. Based on this perspective, Rawls further discusses how to treat the marginalized population.

Norman Daniels extends Rawls's theory in the field of bioethics. He argues for a just health care system that might be based on a Rawlsian principle of "fair equality of opportunity". He considers some properties, such as race, health, IQ, national origin, and social status that often have served as bases of distribution. When the public uses the principle of "fair equality of opportunity", they are often thinking about these properties. This thinking leads to rules such as "To each according to gender" or "To each according to social status". Just this thinking forms a widely accepted reason for permitting different treatment of some persons. To answer this thinking, Norman Daniels argues that some properties, such as IQ and health, are not the responsibility of the people who own them, so they should have fair opportunity to be treated justly (Beauchamp et al., 1994, pp. 334-343). In a similar vein, some philosophers argue that some properties should not be taken as a justification for discrimination, because the persons who have these properties (for example, the homosexual's sexual preference, the drug user's addiction, and the HIV-positive's disease) are not responsible for them.

However, some might say that some HIV-positive and AIDS patients who got the disease through morally improper behavior are responsibile for having their disease, so they do not have the right to be treated justly. I think there is something wrong with this view.

From the point of view of tolerance, such HIV-positive and AIDS patients do not bear all the responsibility for their behavior. AIDS is a special disease, which can be contracted in many ways. In China, some people might be infected by casual sex, having no other way to satisfy sexual desire. Some young people might be infected by premarital and extramarital sex because of the traditional culture's objection to premarital and extramarital sex as well as to making condoms available to young, unmarried people. Some homosexuals, drug users, and prostitutes might be infected as a result of having multiple partners because the repressive climate made them go underground where no HIV education and prevention was available. Even though someone may be infected by unhealthy sex, we have to forgive them, try to understand them, give them a chance and help them correct their mistakes.

Some epistemological theories can justify the deeper reason for tolerance. The limits of human cognition and the lack of a final answer to any ethical question has brought about different points of view. Tolerance is the principle following epistemology and obeying the natural law. The implication of the principle of tolerance for HIV prevention is that we ought to form a lenient moral and legal climate for the work of prevention and education, uniting all groups of the population, making all groups of population assume their duty and responsibility to fight against the common enemy—HIV/AIDS.

Someone may argue that tolerance is the last way to solve ethical issues, so locating the principle of tolerance in first order of a new bioethical framework is wrong. The starting point of this view is that various moral communities must obey one ethical view and be forced to be the same. When a moral community loses its power, it has to apply tolerance to deal with the differences between itself and others. This is a negative tolerance. If the starting point is that the existence of moral difference is a fact in world, we can locate the principle of tolerance in any order which depends on the nature and requirement of that ethical problem. This is a positive tolerance.

Some may argue that we cannot tolerate some particular political conviction or view which objects to the view of the government. However, from history we know that if a government does not tolerate the differences between moral communities, it might lose its power in the end.

The principle of tolerance is compatible with the principle of beneficence and autonomy in my bioethical framework, but it seems to have some incompatibility with the principle of care; we will discuss this problem later.

4. PRINCIPLE OF CARE

The principle of care is located in the last order of this bioethical framework. It improves on the principles of nonmalficence, beneficence and respect for autonomy and helps to solve the issues within the basic bioethical framework, especially about how to solve special issues that HIV/AIDS-related population is involved in.

The principles of nonmaleficence and beneficence are based on utilitarianism and Kantianism. Utilitarianism is a consequentialist theory, which is a theory holding that actions are right or wrong according to the balance of their good and bad consequences. The right act in any circumstance is the one that produces the best overall result, as determined from an impersonal perspective that gives equal weight to the interests of each affected party. Utilitarianism accepts only one basic principle of ethics: the principle of utility, which asserts that we ought always to produce the maximal balance of positive value over negative value or disvalue (Scheffler, 1988, P. 17). Using this principle in medical practice, health care policy makers, physicians and other related persons try to balance benefits, costs and risks.

When applied in an ordinary medical context, this principle is relatively unproblematic, because the conflict between patients and other related persons and society is not so serious. However, the HIV/AIDS epidemic is not the same as ordinary diseases. It is a fatal disease with no effective therapeutic drugs. It is an infectious disease which can be communicated by HIV-positive individuals who live without symptoms. On one hand, HIV infection-causing private acts can have social consequences. If we do not control the behavior of HIV/AIDS people, the disease will produce devastating effects on society. On the other hand, HIV-positive people can live more than ten years without symptoms, and ought to have any rights (such as the right to marriage, work, or education) possessed by other peoples. But when they get married, they might infect their partner through sex action. So the possible conflicts between society, HIV/AIDS individuals and other related people are much larger than ordinary diseases.

To solve these serious conflicts between persons or between a person and society, utilitarians have said that by giving equal weight to the interests each affected party, we can get the best overall result. However, in practice this is equal to permitting the interests of the majority to override the rights of minorities. For the interests of others, HIV/AIDS-affected people might have to give up the right to marriage or having a family. In other words, they may have an obligation to the health of others and to society. Is this demand too much? This is unjust. In this view, utilitarianism fails in the balancing practice of HIV/AIDS prevention.

Kantianism is also inadequate. Kantianism is an obligation-based theory. For Kant, one must act not only in accordance with, but also for the sake of, obligation. That is, to have moral worth, a person's motive for acting must come from recognition that he or she intends to do what is morally required. (Veatch, 1981, p.233) But Kant has a problem with conflicting obligations. Sometimes we cannot at the same time fulfill obligations to two persons. Kant's categorical imperative is both obscure and difficult to render functional in the moral life.

Although universalizability is a necessary condition of ethical judgments, rules, and principles, sometimes it is difficult to solve problems concerning value or interest conflicts in special context, especially in cases concerning, for example, an HIV/AIDS person's marriage or family.

Conflicts of interest are also conflicts of principles; the conflict between an HIV positive person and a health partner wanting to be married is also the conflict between the principles of beneficence and nonmaleficence.

To solve the conflict between principles, W. D. Ross has defended several basic and irreducible moral principles that express *prima facie* obligations. He lists obligations of self-improvement, justice, beneficence, and nonmaleficence, and holds that the principle of nonmaleficence takes precedence over the principle of beneficence when the two come into conflict (Beauchamp *et al.*, 1994, p. 45). But just as Ross says that we must examine the situation carefully when two or more obligations conflict, in some cases, the judgment is about the *weight* of the principles, which is not given by the theory. So Ross's theory also can not solve the balancing problem well.

In China, when meeting the balancing problems, government officials or scholars usually apply the collectivist theory. The proponents of the collectivist theory think everything fundamental in ethics derives from communal values, the common good, social goals and cooperative virtues. So when using the collectivist approach to balance the conflict of interest between an individual and community or society, collectivism's first inquiry is not about the individual's rights, but rather about the community's values. Collectivists would likely ask individuals to give up private rights in order to promote communal values. If someone does not do things like this, he might be regarded as individualist or selfish and get a bad reputation.

If we apply the Chinese collectivism in the HIV/AIDS situation, the HIV-positive seems to have to give up his or her own right to marriage and family. But this conclusion is obviously wrong.

The principle of respect for autonomy is still not suitable in balancing the HIV/AIDS problem, especially in China. Two philosophers have used their own theories to interpret the principle of respect for autonomy. They are Immanuel Kant and John Stuart Mill. Kant argued that respect for autonomy flows from the recognition that all persons have an unconditional worth, each having the capacity to determine his or her own destiny. For Kant, autonomous persons are ends in themselves capable of determining their destiny. Mill argued that citizens should be permitted to develop according to their personal convictions as long as they do not interfere with the expression of freedom by others. (Beauchamp *et al.*, 1994, p. 125)

These thinkers paint a picture of isolated persons living in society with isolated autonomy. In Chinese thinking, individuals are never considered as separate entities, they are always regarded as part of a network, each with a specific role in relation to others. This view came from traditional Chinese moral philosophy, in particular, Confucianism, a mainstream philosophical tradition that shapes Chinese culture and society. Confucianism includes the concept of the relational self, the connectedness of persons and related autonomy. (Yu, 1996, pp. 3-4) The Chinese always decides things by family; this is not same as Western people who promote individual autonomy. So the Western principle of respect for autonomy is not suitable for the Chinese. Most Chinese have only known the concept of duty, but not the concept of right. As in a social relationship, one should respect and value the other party rather than just oneself. Is there a danger of losing the rights of individuals, especially marginalized individuals, in such a framework?

To improve the basic framework and solve the HIV/AIDS related balancing problems, I add a principle called the principle of care. The meaning of the principle care is that from the starting point of caring, one analyses ethical affairs depending on the context and the relationship between parties, and tries to arrive at a conclusion that encourages each party to care for the interests of others so as to make the conflict smaller. The foundational theory of the principle of care is feminist ethics of care.

Ethics of care shares some premises with Western communitarian ethics, including some objections to central features of liberalism, an emphasis on traits valued in intimate personal relationships, such as sympathy, compassion, fidelity, discernment, and love, and a minimization of Kantian universal rules and utilitarian calculations.

The origin of the ethic of care is feminist perspectives. Caring distinguishes between caring about and caring for. "Caring about" something can distance the agent from the object of caring and involves impersonality, cause, institution etc. "Caring for" focuses on emotions, feelings, and attitudes (Holmes & Purdy, 1992, p.118). Central to the ethics of care are the notions of receptivity, relatedness, and responsiveness. Ethical caring is simply the relation in which we meet another morally. Motivated by the ideal of caring in which we are partners in human relationships, we are guided not by ethical principles but by the strength of the caring. In this point, morality is viewed in terms of responsibilities of care deriving from attachments to others, and empathic association is also stressed as the emotional relation with others. Ethical judgement is not based on the primacy and universality of individual rights, but rather on a very strong sense of being responsible.

Another important feature of the ethics of care is its contextual approach to ethical problems. Ethics of care stresses approaching ethical dilemmas in a contextualized, narrative way that looks for resolution in particular details and that involves personalized and socialized contexts. In this way the affairs of ethics can be better understood (Potter, 1993, p.245).

The principle of care also is a new way of reasoning ethically. It does not stress isolated autonomy, avoiding direct conflicts between individual and individual, and focusing on how to be beneficent within the relationship and context.

When we apply the principle of care to solve the problems facing HIV/AIDS persons, the ethical issues are sometimes changed to relationship issues. Relationship issues do not involve deciding who wins various right struggles. There can be more than one conclusion to a rights struggle. For example, from the perspective of caring, we can care both for HIV/AIDS persons and their sex partners by giving rights to marriage and reproduction to both, but whether they can get married or not depends on their relationship, their health, and their related network and so on. In this way, we can know how to assign rights and who ought to receive the rights. This reasoning involves the analysis of relationship, context and consequence. From these features we can get different conclusions for different HIV/AIDS-involved persons in different conditions. Such a conclusion can avoid some conflicts.

Some feminists have argued that utilitarianism, Kantianism, Rawls's justice theory and so on can be called "ethics of justice" compared to "ethics of care". Held goes further, to promote the "ethics of care" and reject the "ethics of justice" (Held, 1995, pp. 1-7) I disagree with the last opinion. Ethics of care need not be hostile to principles. In practice, caring should sometimes be principle-guided. In caring, at times our actions may be too partial and in need of correction by impartial principles like those suggested by Kantians and utilitarians. The ethics of care can be viewed as a supplement to the ethics of justice in balancing conflicts of rights or interests, and the ethics of justice is a rational addition to the ethics of care. From this view, even though there are some incompatibilities in the foundational theories between the principles of beneficence, autonomy, tolerance and care, there are no inconsistencies between principlism and contextualism in my new bioethical framework.

5. CONCLUSION

The new bioethical framework that consists of the principles of tolerance, beneficence, autonomy and care emphasizes both principle and experience, right and responsibility, reason and passion, and individual and community. Even though the compatibility between old principles and new principles or between the ethics of care and the ethics of justice is still not very clear in the field of bioethics, I hope this new bioethical framework can be useful and can improved by further inquiry and application. Finally, I must say that my bioethical framework is formulated only for the evaluation of the Chinese policy of HIV-prevention and deals with HIV prevention problems in China as well as other parts of the world, but it may not be a general bioethical framework to handle every ethical issue in biomedicine.

Department of Ethics & Center for Applied Ethics
Institute of Philosophy, Chinese Academy of Social Sciences
Beijing, China

REFERENCES

Beizhuan Zhang: 2002, ' Yaoxun', *Friend Exchange* 6: 1, Qingdao: Qingdao Medical University Press.
Beauchamp, T.L.: 1982, *Philosophical Ethics -- An Introduction to Moral Philosophy*, New York: McGraw-Hill.
Beauchamp, T.L., Childress, J.F.: 1994, *Principles of Biomedical Ethics*, 4th edition, New York: Oxford University Press.
Gillon, R.: 1992, *Philosophical Medical Ethics*, New York: John Wiley & Sons.
Held, V.: 1995, *Justice and Care*, Boulder: Westview Press.
Holmes, H-B, Purdy, L.M.: 1992, *Feminist Perspectives in Medical Ethics*, Indiana: Indiana University Press.
Potter, A.: 1993, *Feminist Epistemologies*, London: Routledge.
Scheffler, S.: 1988, *Consequentialism and Its Critics*, New York: Oxford University Press.
Veatch, R.M.: 1981, *A Theory of Medical Ethics*, New York: Basic Books.

CHAPTER 19

HO-MUN CHAN

JUSTICE IS TO BE FINANCED BEFORE IT IS TO BE DONE: A CONFUCIAN APPROACH TO HONG KONG PUBLIC HEALTH CARE REFORM[1]

This paper first gives an overview of the healthcare financing system in Hong Kong and then briefly explicates the reasons for reforming the existing financing system. It argues that the reform should be guided by the care-based conception of social justice in Confucianism and that a just health care system should guarantee a decent minimal level of service for all as determined by an open and accountable mechanism of rationing. Based on this argument, the paper concludes that the sustainability of the public health care system in Hong Kong should be maintained by strengthening the rationing measures and revamping the fee structure in accordance with the Confucian conception of social justice.

1. HONG KONG HEALTH CARE SYSTEM'S FINANCING AND REFORM

It is a popular belief that the Hong Kong Government has all along practiced the *laissez-fare* policy of economic non-intervention. Such a belief is indeed misconceived. The Government no doubt maintains a modest expenditure pattern in social security and exerts rare intervention in the spheres of capital, commodities and the labor market. However, the Government still performs an active role in making substantial investments in the spheres of housing, education and health care, and is a key service provider in these spheres (Wilding et al., 1997).

Early in July 1974 the Hong Kong Government released the *White Paper on the Further Development of Medical and Health Services in Hong Kong*, which affirmed the policy that no one should be denied proper health care services due to lack of means. At present, the daily charge for hospitalization in the public hospital is HK$68 (around US$8.5) which amounts to 2-3% of the average cost, while, according to the by-census??? of 1996, the monthly median income of an ordinary household (3.3 persons) is HK$17,500. The daily charge for hospitalization at such a low level would not place much of a burden on the great majority of households.

R.-Z. Qiu (Ed.), Bioethics: Asian Perspectives, 207-228.
© *2003 Kluwer Academic Publishers. Printed in the Netherlands.*

Furthermore, the Government could also exempt those with financial difficulties (e.g., those dependant on the comprehensive social security allowance) from the hospitalization charges. The financial accessibility of the Hong Kong public health care system can therefore be said to be very high.

Currently, more than 90% of hospitalization service is provided by the public hospitals (Ho, 1997, p. 28; Tse, 1998). The total health care expenditure only accounts for 4.81% of the territory's GDP while the parallel figure of many developed countries is around 8% (Langan, 1998, p. 39). The public health care expenditure amounts to only 2.16% of the territory's GDP in Hong Kong, even less than the 2.65% of the GDP attributable to the private health care expenditure, but the public health care expenditures in many developed countries amount to around 5% of their GDPs (Langan, 1988, p. 39). Given the relatively low health care expenditure, Hong Kong has nevertheless witnessed the ever-rising quality of its health care services which have indeed attained a remarkably high standard. For example, Hong Kong people's life expectancy and the infant mortality rate are respectively higher and lower than that of the United States (Tse, 1998).

The Hong Kong Government has been effectively monitoring the growth of public health care expenditure. Although the Government has allowed to some extent the growth of public health care expenditure in real terms, growth may only keep pace with that of the territory's GDP. The public health care expenditure has always been kept at approximately 2% of the territory's GDP, representing a fairly constant portion of the total public expenditures as well. This is a result of the Government's endeavor to maintain a rather constant ratio between the public health care expenditure and the total public expenditure. The public health care expenditure has throughout the past ten years experienced constant growth in real terms at an annual rate of 7.8% (Tse, 1998). Such growth is largely attributed to Hong Kong's economic development, and this growth rate is more or less equivalent to that of the territory's GDP. The recurring public expenditure devoted to health grew from 13.1% in 1990/1991 to 14.5% in 1996/1997. In the meantime, the total pubic health care expenditure (including the recurrent expenditure) has been kept at 11-12% of the total government expenditures (Tse, 1998).

However, there are always tensions among financial accessibility, quality of health care services and financial affordability. The extremely high financial accessibility and the ever-increasing quality of health care services necessarily overburden the Hong Kong public health care system given that the health care budget is so limited. The Hong Kong Government often tries to alleviate pressure on the public health care system by raising efficiency and reducing waste. Yet the effectiveness of this effort is strained, particularly due to the aging demography, ever-increasing cost for medical technology and ever-rising public expectation of high quality services (Hong Kong Government, 1993). The public health care expenditure is destined to rise. It is estimated that the public health care expenditure relative to the total government expenditure will rise by 50% in 2016 (Harvard Report, 1999).[2] Due to the possible financial restraint, the Government perceived that growth at such rapid rate would necessarily hinder the development of the other social services.

It is apparent that the objective of the reform should not be confined to alleviating the financial burden of the public health care system. Had that been the sole objective, why would the government not have just charged the user as much as the service costs? Such charging definitely reduces the financial burden of the public health care service system to zero, doesn't it? We would not endorse such charging since many of the users would not be financially able to meet such costs and would then be denied the services, which seems to be an apparent violation of social justice. Indeed, the government has all along stressed the primary principle of its health care policy to be that no one would be denied basic health care services for lack of means (Hong Kong Government, 1974, 1993; SCMP, 8/12/98). The adoption of any reform option is therefore expected to be guided by such a principle.

However, the primary principle advocated by the Hong Kong health care policy is still obscure—that is, what amounts to a basic level of health care services? In this paper, I'll argue that this principle indeed agrees with the principle of social justice and strives to articulate a decent minimum of health care as the basic level from the Confucian perspective of social justice. I also discuss how such a decent minimal level could be determined by an open and accountable rationing mechanism.

2. SOCIAL JUSTICE AND THE DISTRIBUTION OF HEALTH CARE BURDENS AND BENEFITS

We may interpret social justice as follows: social justice commands us to be committed to treating equals equally, which means that everyone should be entitled to one's due, i.e., two persons who are equal in the relevant respects should shoulder equal social responsibility and receive equal social goods. However, in the absence of some theoretical precepts, such an interpretation could hardly afford us substantial guidance, since we are still left with questions like: by what criteria can two persons' relevant respects be regarded as equal? How is one to determine the extent of a man's social responsibility to be borne and social goods to be received? (Miller, 1976). For instance, regarding the responsibility and distribution of health care services, should we consider financial affordability as the criterion in determining if two persons are equal? Should we allow those of means to pay more and receive better service? If we are prepared to accept treating health care services as commodities, our answer to the two aforementioned questions should be in the positive. But many poor people would not be able to afford sufficient health care service in the market. Would the commodification of health care services turn its back on social justice?

2.1. Libertarianism

Libertarians hold that social goods and services should be allocated by the market mechanism and the costs for the provision of such goods and services should be borne by the user. They therefore support the commodification of social goods and services and oppose any state intervention of levying taxes to finance social goods and services which eventually are distributed to the user at a very modest charge or even for free. Such state intervention cause the rich, who pay more tax, to shoulder more but not necessarily receive more and better services, while the lower-income groups are not required to pay tax but are able to receive free or almost free health care services. Libertarians regard such an arrangement as "robbing" the rich to benefit the poor, infringing on the right to private property, and compelling the higher-income groups to subsidize the lower-income ones. Libertarians believe that in such a situation, the higher-income groups would therefore not be entitled to all their labour products. Their labour-products, which go to the tax revenue, could be regarded as forced labour, and that is contrary to social justice (Nozick, 1974).

Libertarians recognize that there is one and only one criterion for the notion of social justice, which means that all social goods and services are to be commodified. Their costs are to be borne by the user and their distributions are to be effected by the user's financial affordability. Libertarians reject any other criteria, such as needs, since the market mechanism may not satisfy all needs. In the event of such market failure, the state is lured to exert intervention as a corrective action. Libertarians again think of such intervention as contrary to social justice.

It is obvious that libertarians would advocate the commodification of health care services, which indicates that the costs should be borne by the users via insurance or direct payment (Graham, 1988). The criterion for the distribution of health care services is thus its affordability to users. The libertarian justifies this stance in various ways. Firstly, they take the view that one should be responsible for one's health. According to them, many diseases are unrelated to hereditary factors, but are more connected with one's living style, compliance with a doctor's prescription, engaging in high-risk activities and so on. If the health care services are to be distributed on the basis of needs, those who pay attention to their health would necessarily be compelled to subsidize those with less diligence.

Secondly, a health care system entirely dependent on government funding would compromise the citizen's options for health care services. Some may not be too concerned with their health and would only prefer to spend a limited sum on health care services. Others may be keen to contribute much more to have health care services of impressive quality. They may find the public health care service as not up to their expectation and turn to a private health care service provider. Since they may have already paid their portion of tax which goes to finance the public health care services, they de facto have paid twice when they do not seek services from the public sector but go to the private sector instead. Quite the contrary, those with less consciousness regarding their health may find themselves being driven to pay too high a price for a health care system of excessive quality. They are therefore deprived of the option of lessening their medical expenses.

Thirdly, libertarians argue that taxation would cause two kinds of inequality. In the first place, the higher-income groups pay more than the lower-income groups do but the health care services they receive are the same. The higher-income groups are therefore forced to subsidize the lower-income ones. In terms of absolute figures, the lower-income groups pay less tax but the tax they pay may indicate higher marginal cost as their income basis is relatively less. If the government is prepared to levy less tax by commodifying the health care service, those lower-income groups may retain more disposable income for non-health care purposes because they are no longer compelled to spend a fixed amount of their money on health care through taxation. Now, with less contribution from those lower-income groups, it would cost more for the higher-income groups in terms of medical insurance to enjoy the same level of health care service as before commodification. So we can say that a public health care system may compel via taxation the lower-income groups to subsidize the higher-income ones so that they can enjoy a level of service which is better than that obtainable in the free market.

Libertarians do not stand against (and may even endorse) lending a hand to the financially disadvantaged so that they may receive proper health care service. However, their primary concern is freedom of choice, which means that any helping hand should be lent willingly, not out of obligation. At this juncture, we may be aware of the adverse consequences of viewing freedom of choice as an absolute end in the sphere of health care—the financially disadvantaged group would not be able to afford in the market the basic health care service that they need, and they may not be able to acquire sufficient health care service merely through charity either.

2.2. Egalitarianism

Egalitarians believe that freedom of choice should not be treated as absolute, and that libertarianism has not taken the health care needs of the worse-off seriously. They maintain that one's life is largely shaped by forces beyond one's control (Rawls, 1971). It would not be just to expect everyone to face all ills of life, such as poverty and poor health, on their own. One's poor health is attributed to all sorts of causes, including infection, accident, heredity, aging and so on, which are not within one's control. Even if one's poor health is constantly related to his habits and choices (like being an alcoholic or engaging in high-risk sports), such habits and choices are often caused by one's social background and chance events. The sick may not be competent to take care of themselves and it is thus unjust for them to fight off alone the mischief of poor health, which is a result of natural and social lotteries.

Nevertheless there is still a question: How much medical care should be provided for the sick? Radical egalitarians advocate that everyone's medical demand should be accommodated and they prefer a public health care system which would not enable the rich to pay more for better and faster services (Nielsen, 1989). Such a radical egalitarian stance has ignored the fact that the costs of health care service for many serious sicknesses are so high that society could not afford to satisfy all such demands. On the other hand, we should not presume that the distribution of health care service is a zero-sum game. Someone's receipt of better health care services in the market does not lead to any decrease in quality of service in the public health care system.

Some moderate egalitarians endeavour to reconcile individual freedom and the demand for proper health care service. They invoke Rawl's theory of justice to set down the level of health care service which should be borne by the society. Briefly, Rawls (1971) stipulates that a just society should satisfy the following principles:

(1) Each person is to have an equal right to the most extensive total system of equal basic liberties compatible with a similar system of liberty for all;
(2) Each person is to have fair equality of opportunity;
(3) The distribution of social and economic goods must cause the socially worse-off to live in the best condition (Difference Principle).

According to principle (1), there is no reason why society should not allow people to seek satisfying health care service via the market. But the society has to uphold fair equal opportunity while the aforesaid principle is to be observed. Norman Daniels (1981) thought that, for the sake of fair equal opportunity, everyone should have access to a normal range of opportunity so as to steer one's life. Whether a person is able to maintain a normal range of opportunities depends on whether he is able to possess the normal species functioning of a man as homo sapiens. Daniels accepted that the resources available for the health care service of a society is scare and the government is therefore justified in maintaining just the normal species functioning of every citizen.

Daniels wanted to have a society which could ensure that everyone had access to a decent minimum of health care. The idea of a decent minimum provides a moral justification for a two-tiered health care system: A public health care system or any other social security mechanism which is to satisfy everyone's demand of a decent minimum of health care; a private health care system which allows people via the marketplace to pursue better services by paying more. Such an arrangement appears to reconcile individual liberty and the basic (i.e., decent minimal level, hereandafter) health care demand (Daniels, 1981; Buchanan, 1984; President's Commission, 1983; Beauchamp & Childress, 1994 Ch 6).

The concepts of "two-tiered system" and "a decent minimum" as advocated by Daniels are appealing. What is controversial is his setting "a decent minimum" at the level which ensures one's normal species functioning. This may be too demanding for the following two reasons. First, a poor society may not have sufficient resources to provide services up to such a level and this could not be regarded as unjust. Second, to restore the chronic sick person with his normal species functioning may not be financially viable or even not medically possible and therefore even a wealthy society may be not able to achieve such a level.

Another way to attain a decent minimum of health care is by invoking Rawls's Difference Principle. The Difference Principle stresses that a society should optimize the socially worse-off to live in the best condition. Having said that, society is not compelled to shoulder alarming burdens. For instance, the Difference Principle can only require an economically backward society to provide very limited health care service free for the poor simply because any further attempt to raise the health care service standard would necessarily hamper economic development and those worse-off would in turn be made even worse off (Campbell et al., 1997, Ch 13). In a more economically advanced society, the Difference Principle would divert more resources from the upper and middle classes towards the lower class as it pursues the goal that the worse-off could live in the best possible situation, rather than in the condition of a decent minimum. The upper and middle classes thus have grievances as to their excessive contributions. A number of psychological studies revealed that when people are placed in an experimental setting which simulates the effect of being in the original position behind the veil of ignorance, they only prefer to have a society where the living condition of a decent minimum can be guaranteed for all. (Frohlich & Oppenheiner, 1992). Recently, I have conducted a similar study in Beijing, Taipei and Hong Kong verifying this point.[3]

2.3. Public Choice Theory and Socialized Care

It is no easy task to provide a morally adequate characterization of a decent minimal standard of health care. Rights-based egalitarianism always starts with the preferred state of individual health. Whenever a person's health deviates from what is normal, he or she suffers and there is a need to have such a deviant health state normalized. A person's such need, according to egalitarianism, becomes one's right which the government bears the duty to satisfy (Daniels, 1981). However, as pointed out earlier, to cure someone's sickness could be very expensive, quite beyond the capability of a government.

Libertarians reject medical need as a right, and this view has some merit. They distinguish the "unlucky" from the "unjust" (Englehardt, 1981). A person by his bad luck contracts a disease, and other people need not be responsible for that. Contracting a disease is unlike a right being infringed, as any infringement of right is caused by other's mischief. Everyone has the duty to respect the personal freedom of others and not to hurt them. When anyone infringes another's right, society is justified in holding that wrongdoer to be responsible for any resulting loss and suffering. It is important that everyone's fundamental rights should receive equal respect and we are not allowed to choose to disregard one person's right being infringed but not another's. The infringement of one's rights is a kind of unjust treatment. But sickness is always attributed to some natural and social factors which are beyond one's intent and control, like the cases of natural disaster and accident. It is only proper for us to say that being sick is unfortunate but no others are to be held responsible for that. This situation should be distinguished from a case in which someone's right is infringed. If we are keen to regard the sick person's medical needs as 'rights', such 'rights' could only be 'manifest rights' or 'imperfect rights'. It is desirable for us to lend a hand to those less lucky, but the mere fact that we are less altruistic does not render us morally wrong. Even if we choose to help one but not the other, we could hardly be blamed for offering unequal treatment. Having said that, we do not mean to be indifferent to another's suffering. It is only stressed that the morality of being altruistic is not as strong as that of respecting others' fundamental rights. We therefore should not treat a person's medical need as a right (President's Commission, 1983).[4] It should never be thought that caring for the sick is trivial. Even libertarians accept that caring for the sick has a significant moral claim, but they only accept this to the extent that such caring is the product of voluntariness and is affordable to the society. The libertarian view should not be embraced uncritically. The following scenario is likely to haunt us if we are to care for the sick, and the financing of such medical care is not without difficulty (Buchanan, 1984; see also Sunstein, 1991):

Suppose that a person contemplates donating money for the society's health care services. Impressed by the possible enormous extent of such expenditure incurred, he may think: "If the society has had enough for health care, my donation is better to be diverted elsewhere. On the contrary, if the society does not have enough, my donation is just another tiny contribution to the purported massive sum, making no difference at all. My money should go somewhere else." When more and more people harbor such a view, society would never attract enough money for health care services. People tend to think that, without the assurance of others' contributions, their contribution would just be too negligible. It follows that the society is unable to amass the required sum via people's altruism.

The above situation stands as a classic example of the coordination problem according to public choice theory. In our daily lives there are plenty of such examples. For instance, many people would prefer to drive without rules and regulations, but that preference would only lead to traffic collisions. Hence, we cannot resist every attempt of the state intervention, since taking individual freedom as absolute is nothing but a mess. An effective measure to avoid such a mess is to regulate traffic by the promulgation of relevant laws requiring, for example, that automobiles drive on the left of the right. Such a suggestion by no means implies that state intervention is always desirable. For instance, we would not endorse similar control measures on pedestrians unless there appeared to be a security reason for crowd control (like on the occasions of large celebration). The state intervention is deemed to be unnecessary if agreements can be reached among people themselves.

The difficulty in financing the health care service lies not in the restrained capability of society and the unwillingness of the people to help the sick. If the coordination problem went on without getting solved, society could not finance its health care system via voluntary donations in spite of the presence of the ability and good intentions of society. It is suggested that only the state intervention of taxation and other mandatory contribution schemes could solve such a coordination problem. Due to the peculiarity of medical need and services, libertarians should accept that the moral claim to care for the sick is justified to be an "enforced beneficence" so long as they accept the existence of such a moral claim. Otherwise, the moral claim to care for the sick becomes empty.

The above proposal avoids the hard task arising from the determination of a decent minimum of health care as advocated by the egalitarians. Such a hard task is due to the egalitarians' sole concern for the need of the sick without proper regard for the cost impact on the whole society or different income groups. Our reasoning could eliminate the aforesaid problems. On the one hand, we, unlike libertarians, care about the sick as egalitarians do. Yet we can avoid the problem of overburdening that egalitarians need to confront, because the government is required to provide only a basic level of health care that the public can afford. On the other hand, we admire individual freedom as libertarians do, but do not treat it as absolute. Whenever there looms the coordination problem, we accept state intervention to ensure the availability of a basic level of health care service. Although we prefer the individual acquisition of health care services via market mechanism, we also suggest a two-tiered system consisting of both public and private elements which would ensure that no one was deprived of a basic level of health care due to his lack of means. We believe that such a proposal will enable us to have the best of the two worlds of egalitarianism and libertarianism.

Although the above position carries its unique merit by basing the social justice of the health care system on care instead of individual rights, one may challenge the above view by arguing that there is an unwarranted assumption about the value of care and its relationship with justice. A libertarian may argue that if members of a society in general do not treasure the value of care, the coordination problem that we have examined does not exist because no beneficence needs to be enforced. So the public choice theory alone cannot provide adequate support to the above position. It has to be strengthened by a public ethic of care. As to the relationship between justice and care, some feminists tend to think the two are incompatible (Gilligan, 1982) but others think just otherwise (Freidman, 1987; Slote, 1998). The latter view seems more applicable to the problem of justice in health care service because the problem is about who is to care and who is to be cared for. Justice and care should not stand apart, and whenever basic health care becomes unavailable due to the state's lack of coordination, injustice stands out. I believe that this position can be justified by a Confucian conception of social justice founded on a distinct ethic of care.

3. CONFUCIAN CONCEPTION OF SOCIAL JUSTICE

Social justice is of primary value in the political philosophy of Confucianism. In the *Analects*, Confucius said,

> [The head of a state or a noble family] is not concerned lest his people should be poor,
> But only lest what they have should be ill-apportioned.
> He is not concerned lest they should be few,
> But only lest they should be divided against one another (*Analects*, 16.1).

He believed that if all is well-apportioned, there will be no poverty and society will be in harmony and stability (*Analects*, 16.1). So there is some similarity between Confucius and Rawls in the sense of their both believing that justice is essential for underpinning a well-ordered society.

3.1. The Doctrine of Benevolent Government

In the Confucian tradition, it is alleged that in order to maintain justice, the ruler of a state should follow the doctrine of benevolent (ren) government. Though Mencius is a well-known proponent of this doctrine (Hsu, 1975; Li, 1999), Confucius had already stated the idea in the *Analects*: "Zigong asked about government. The Master said, "Sufficient food, sufficient weapons, and the confidence of the common people" (Analects 12.7). To practice the governance of benevolence (*ren*) is to look after the well-being of the common people. If a ruler can take good care of the people, they will trust him and he can then maintain his political legitimacy and the trust of his people; so being caring is the way of becoming a wise ruler. This idea was further developed by Mencius:

> ...a wise ruler will decide on such a plan for the people's means of support as to make sure that they can support their parents as well as their wives and children, and that they have enough food in good years, and are saved from starvation in bad (Mencius 1, 7).

> If your Majesty wants to run a benevolent government, why not turn to what is of fundamental importance? Let mulberry trees be planted about each homestead to five *mu* of land, and those who are fifty will have floss silk garments to wear. Let fowls, pigs and dogs be raised without neglecting their breeding seasons, and those who are seventy will have meat to eat. Let farm work be done without interference in a hundred *mu* of land, and a family of eight mouths will not go hungry. Let careful attention be paid to education in local schools, where the significance of filial and fraternal duties is stressed repeatedly, and grey haired people will not be carrying loads on the roads. In a state where old people are clothed in floss silk garments and have meat to eat, and the masses do not suffer from hunger and cold, what prince can fail to unify the whole world? (Mencius 1, 7).

Mencius had seen people being taken away by the ruler during farming seasons, making them unable to support their families, and he believed that taking good care of the common people was essential to good governance and political legitimacy; a ruler who cares about his people can gain legitimacy by popular support if he can ensure that people do not miss their farming seasons, have good harvests in good years and are prepared for bad years. On the contrary, a despotic ruler who does not care about the well-being of his people loses legitimacy and Mencius believed that "there is duty not just to oppose but also to *depose*, any unjust ruler" (Hall and Ames, 1999, p.171).

Although the idea of a democratic form of government cannot be found in Confucianism, there are nevertheless some quasi-formulations, elements or seeds of democracy in the doctrine of *minben* (people-as-basis). In the *Book of Documents*, it is written: "The masses ought to be cherished, not oppressed, for it is only the masses who are the root of the state, and where this root is firm, the state will be stable" (Legge, 1960, Vol. 3: 158). In Mencius, we read, "Of the first importance are the people, next comes the good of land and grains, and of the least importance is the ruler. Therefore whoever enjoys the trust of the people will be emperor" (Mencius 14, 14). So in the Confucian tradition, cherishing the common people and allowing them to flourish is of utmost importance in good governance. But the idea of *minben* (people-as-basis) cannot be equated with that of a full-fledged democracy. According to Ambrose Y. C. King, "[*Minben*] contains the essence, perhaps, of what Lincoln meant by 'of the people' and 'for the people', but it does not extend to 'by the people'" (King, 1997, p. 172).

3.2. The Confucian Ethic of Care

The philosophical foundation of the doctrine of benevolent government and the idea of *minben* can be traced to the care ethic in Confucianism. Such an ethic can be derived from the notion of benevolence (*ren*). According to Chan (1993), the concept of *ren* has two senses. As a particular virtue, it refers to the virtue of benevolence and altruistic concern for others. As a general virtue, *ren* stands for perfect virtue, goodness or moral perfection. *Ren* in the general sense provides the overarching, unifying ethical framework in Confucian ethics; it encompasses all the other particular virtues such as benevolence, propriety, courage, filial piety, loyalty, etc., and so is the virtue of virtues. Li (1999) makes a similar distinction: *ren* as affection vs. *ren* of virtue. As an affection, *ren* "stands for the tender aspect of human feelings and an altruistic concern for others" (p. 96). In the second sense, a person of *ren* is a morally perfect person and the virtue of *ren* can only be realized "among other virtues" (p. 97).

The first sense of benvolence (*ren*) is actually captured in the notion of care found in the work of Confucius and Mencius:

> "Fan Chi asked about the benelovent. The Master said, "He loves men" (*Analects* 12, 22).

> "A young man's duty is... to have kindly feelings towards everyone..." (*Analects* 1, 6).

> "A benevolent man loves everybody" (*Mencius* 13, 46).

> "A man of benevolenc loves others" (*Mencius* 8, 28).

The same usage of the notion of benevolence (*ren*) can also be found in the writings of other ancient Chinese philosophers (Tao, 1998).

According to Mencius, the source of this first sense of *ren* can be traced to the natural heart that is sensitive to the sufferings of others. Mencius wrote,

> All men have a sense of compassion... The reason why I say all men have a sense of compassion is that, even today, if one chances to see a little child about to fall into a well, one will be shocked, and moved to compass on, neither because he wants to make friends with the child's parents, nor because he hates to hear the cry of the child... The sense of compassion is the beginning of benevolence. (*Mencius* 3, 6).

So, the ethic of care and compassion occupies a cardinal place in the Confucian notion of benevolence (*ren*). However, it is a mistake to interpret the philosophy of *ren* as a philosophy of universal love, which means that one ought to care for everyone equally. Both Confucius and Mencius advocate that the practice of *ren* should start within the family:

> "Filial piety and brotherly respect are the root of benevolence" (*Analects* 1, 2).

> "It is benevolence to love one's parents" (*Mencius* 13, 15).

Family ethics are central to Confucianism and this makes some scholars such as Bertrand Russell uphold the mistaken view that filial piety is the weakest point of Confucian ethics because it prevents the growth of public spirit (Russell 1922). Confucianism in fact advocates that people should extend the practice of *ren* from within the family to other people. Mencius is well known for this principle of "care by extension". The following are his formulations of the principle:

> Do reverence to the elders in your own family and extend it to those in other families; showing loving care to the young in your own family and extend it to those in other families (*Mencius* 1, 7).

> A benevolent man extends his love from those dear to him to those he does not love (*Mencius* 14, 1).

The extension of love to those who are not related to you does not mean that you should love your father and a stranger equally. Your concern about the well-being of a stranger is only an extended love and so it should be weaker than your love for your father. The extended love should be guided by the principle of "love with gradations" formulated in the following passages:

> A gentleman treats all things carefully but there is no benevolence shown in his attitude towards them. To the people he is benevolent but not affectionate. A gentleman is affectionate to his parents and relatives, so he is benevolent to the people. He is benevolent to the people, so he is careful with things. (*Mencius* 13.45).

> A benevolent man loves everybody, yet his relatives and the virtuous should be given first place in his heart. (*Mencius* 13.46)

So, according to Mencius, the principles of love with gradations and extension of care are indeed two sides of the same coin.

These two principles can provide a coherent framework for resolving a dilemma in the feminist debate about the role of care in public life (White, 2000). On the one hand, some feminists such as Nel Noddings (1984) believe that the duty to care is limited and delimited by relation. But upholding this view leads one to go down the

road of parochialism, which denies the existence of any caring relationship among
distant people and makes it difficult to answer the objection from libertarianism to
the thesis that the public provision of care is incompatible with justice. As a result,
non-related beings become moral strangers, the ethos of care is eroded and the
public domain regulated by the minimalist ethic advocated by the libertarian. On the
other hand, some feminists such as Susan Okin (1989) go to the other extreme and
maintain that Rawls' egalitarian liberalism "is a voice of responsibility, care, and
concern for others." Yet this sounds psychologically implausible because people in
general can hardly have such strong motives to maximize the well-being of the
worse off who are likely to be strangers to them.

In a culture dominated by the Confucian tradition, the true picture is likely to fall
between these two extremes. The principles of care by extension and love with
gradations are fundamental to the care-based conception of justice in Confucianism.
The two principles imply that the commitment to the care for distant others will fall
somewhere between that of egalitarian liberalism and libertarianism, and so such a
care-based approach to the problem of justice can avoid the difficulties that
egalitarianism and libertarianism cannot overcome. In a Confucian society people
will on the one hand support the government's strong commitment to a caring
society, but on the other hand maintain that the commitment should not lead to the
society becoming a welfare state. The use of the two principles on the one hand
manifests the strong commitment to care, but on the other hand also allows room for
the Confucian values of meritocracy, work ethic, self-reliance and individual
responsibility to flourish and unfold themselves in the free market.

These values occupy a central role in the Confucian tradition and can be traced to
the doctrine of equal moral worth. *Mencius* wrote:

> "...all men may become Yaos and Shuns" (Mencius 12, 2),

where Yao and Shun are sages in Ancient China. In the *Analects*, Confucius
remarked:

> "By nature, men are alike; by practice, they become very different"
> (*Analect* 17, 2).

In the Confucian tradition, everyone has an equal potential and obligation to strive
for good, and so the unfortunate should be taken care of so as to enable them to have
the opportunity to strive for perfection. But good can only be achieved through self-
cultivation and self-realization and that is why Confucianism upholds the values of
meritocracy, work ethic, self-reliance and individual responsibility. So people in a
Confucian society will generally give much stronger support to the notion of
equality of opportunity than to that of equality of conditions or outcome.

3.3. Implications for Social Policy Making

Social policy in Hong Kong has been guided by the Confucian conception of social
justice. The Government has consistently made the obligation to provide care for the
needy a central element in the development of all major social policy. In housing,
the government had already asserted by the early Seventies that public responsibility
was necessary because "the inadequacy and insecurity of housing" and the harsh
situations that result from it, "offends alike our humanity, our civic pride and our

political good sense" (Hong Kong Government, 1972); after years of development, more than 50% of the population now lives on public housing estates. In education, the government accepts "a special responsibility to ensure that no one is deprived of a place in the education system because of a lack of means" (Hong Kong Government, 1994). Although only a very small number of primary and secondary schools are directly operated by the government, the majority, which are aided schools run by religious bodies, voluntary agencies, welfare organizations and trade associations, are funded and fully supported by the government; only very few schools are privately funded. In social security, the government's policy undertakes the responsibility to help "the disadvantaged members to attain an acceptable standard of living" (Hong Kong Government, 1995b).

However, despite a strong commitment to the provision of care, it is not the policy of the government to make Hong Kong a welfare state. As Chris Patten said, "... [a] sense of compassion is the other side to Hong Kong's work ethic. In Hong Kong, no one seriously disputes the need to balance the work ethic of the majority with the welfare requirements of an unfortunate few" (Hong Kong Government, 1995a). Many people in Hong Kong believe that it is important to strike a balance between the values of care, meritocracy, work ethics, self-reliance and individual responsibility. As this paper will argue, this view is rooted in the Confucian tradition. As a manifestation of such a tradition, the social policy of Hong Kong has always aimed to enable people to help themselves by guaranteeing a reliable safety net for all. This view was reasserted in the Tung Chee-hwa, the First Chief Executive of the Hong Kong Special Administrative Region, in the Policy Address of 2001: "Our aim is to help our people to enhance their ability to help themselves and to boost their will-power to do so. Under this social contract, the Government is firmly committed to providing a reliable safety net as a basic guarantee for our citizens" (Hong Kong SAR Government, 2001).

In other words, unlike the situation in a welfare state, the provision of care is maintained at a level that will not lead to erosion of the values of meritocracy, work ethic, self-reliance and individual responsibility. This difference has been consistently proclaimed in government publications, "Hong Kong is not a welfare state but the community cares deeply about the state of welfare" (Hong Kong Government, 1996, 1997; see also Social Welfare Department, 1999). Hence, despite a strong commitment to the vision of care, the government has consistently maintained the level of public expenditure at below 20% of the Gross Domestic Product (GDP) by placing all major policy areas under tight budgetary control. For example, total health care expenditure has been capped at consuming around 11-12% of total public expenditure and 2 to 2.5% of GDP. This control of public expenditure has enabled the government to keep taxation low; less than 40% of the working population are required to pay tax and only 0.3% need to pay the standard top rate of 15%, but their contribution amounts to 19.5% of the total salary tax collected.

In Hong Kong, it is generally believed that economic growth can be best achieved by the free market system, and that the widespread economic prosperity created will enable most people to flourish without relying upon the government. Many people look to the free market rather than to the government for fulfilment of their hopes, but the government is firmly expected to undertake the obligation of taking care of the needy so that they can eventually help themselves.

4. A CONFUCIAN APPROACH TO HEALTH CARE REFORM

As we have seen in the beginning of this paper, the public health care system of Hong Kong has been quite capable of meeting the basic health care needs of the population. However, since both the financial accessibility and the service quality of the system are high, many patients who can afford to pay more may be tempted to use public health care services. For example, the emergency room service had all along been free until late 2002, but many patients who could meet their needs in private out-patient clinics by paying the fees out of their own pockets would rather go to the emergency room to enjoy the free services. The problem had been quite serious even before the economic recession triggered by the Asian Financial Crisis. The values of self-reliance and self-responsibility have therefore been compromised and the public health system has become less sustainable. Furthermore, it is predicted that the aging demography, the ever-increasing cost for medical technology and ever-rising public expectation of high quality services will make the problem of sustainability more severe.

The public health care system in Hong Kong has high financial accessibility because of its adherence to the principle of universal coverage. The public will not welcome the reform of replacing the existing system by a "means-tested" system in which only people with income lower than a certain level are eligible for using public health care services with heavy subsidy from the government. The main reason for rejecting such a reform initiative is that the elderly, the chronically ill and those who need medical services that will pose major financial risks to them may not be able to meet their health care needs in the free market even if they are not from the lower income groups. In particular, the tax payers will have a strong resistance to such a reform because the change will mean that they have to contribute via taxation to a public health care system which offers them no protection against health care risks, especially those that pose major health or financial risks. As Tao has said, if such a new system is in place, "they will be excluded from the care orbit of the government. Their care-giving action will not be reciprocated by care-receiving action. The new structure discriminates against their need for care even though they themselves are also agents of care in the process of care" (Tao, 1999).

A better way to deal with the problem of sustainability is to make use of a rationing mechanism that targets the limited resources to meet the basic needs for all so that the public health care system can provide the universal coverage of a decent minimum of health care. This indeed agrees with the "New Universalism" advocated by the World Health Organization, which means "universal coverage for all, but not coverage for everything" (World Health Organization, 1999). For services that go beyond such a level, either they will not be provided by the system or the patients will be required to pay extra. The introduction of such a rationing mechanism can be justified by the Confucian principles of extension of care and love with gradations because our commitment to help the needy is strong but not unlimited. If this approach is adopted, the next immediate question is how to determine the level of a decent minimum. Prior to answering these questions, we will have a brief account of the different levels of the distribution of health care resources, followed by a deliberation of a mechanism to determine a decent minimum level.

4.1. The Three Levels of Resource Allocation

The distribution of health care resources consists of macro-allocation and micro-allocation; the macro-allocation could further be divided into two levels (Caplan, 1992, Ch 19: Beauchamp & Childress, 1994, Ch 6; Garreet ett al., 1998, Ch 6; Daniels 1987). The macro-allocation of health care resources concerns the following: First, what should be the ratio between the total health care expenditure and the territory's GDP? How much should be borne by the government? Second, what should be the items of service provided by the public health care system? What should not be provided? What is the extent of subsidy rendered for each item? The micro-allocation (the Third Level) concerns what service should be provided for a specific patient. Confronting the restraint that the government is only able to provide limited resources for a specific item (e.g., haemodialysis), the front line medical personnel have to make the clinical decision of who should get the service first.

The relationship among the above three levels is not just top-down, but also bottom-up and interrelated. Due to the rapid changes of social circumstances and the health status, some diseases (like senile dementia and other occupational diseases) could be more prevalent than before. It may also be the case that improved medical technology would cure those patients more effectively, but such technology unsurprisingly costs more than before. Due to the restraint of resources, the front-line medical staff cannot accommodate all patients' needs and they must turn to the third level (the micro-level) to make the hard decisions to delay the satisfaction of those with less imminent medical demands. To alleviate the burden of hard decisions at the third level, the reallocation of resources at the second level could be effected to bring forward more resources to those departments in need. However, the hard fact is that many health care services compete for resources among each other and the scarcity arising could not be solved without having recourse to the first level of distribution. That means the increase in the total expenditure of the public health care services. This, however, may not be attainable as the Government may in the meantime be confronting the ever-rising demands of housing, social security, education and so on. It is thus evident that the first-level distribution necessarily affects the distribution pattern at the other two levels.

At the outset of this article it was said that regarding the first-level distribution of the Hong Kong health care system, the Government is always targeted at maintaining a rather low level of public health care expenditures relative to the territory's GDP and the total government expenditure. Effectively the Government is rationing the health care resources. As such resources are not free from scarcity, the public health care system cannot provide all the required services or make the waiting time tolerable. Those patients with sufficient means may then turn to the private sector for satisfaction while those without will be turned away by the market.

The Hong Kong Government maneuvers to confine the public health care expenditure relative to the territory's GDP. But it is not so evident that the Government devises any concrete mechanism to determine the extent of the public health care services and the share assigned to each category of service. Without such a mechanism, we cannot be sure that no basic service is neglected or that no specific item (like haemodialysis) is supported with too few resources or that the waiting queue for the service in question is too long.

There is a misconception among the people that rationing would lead to denial of health care services. However it is always the truth that, subject to the restraint of resources, the public health care system is indeed conducting *de facto* rationing by assigning resources to different categories of services even without proclaiming a formal rationing mechanism. In the current system, there are always categories of services which are unavailable or available but limited. People therefore have to wait, perhaps for a long time. In the absence of a concrete and formal rationing mechanism, there may not be a set of definite criteria to determine the distribution of resources and the priority of the provision of services. It follows that rich medical resources being injected into certain categories of service does not necessarily represent the considered priority, but is rather the result of bureaucratic politics or just the mere consequence of copying the old pattern. We will then not be able to discern which category of service is deprived of resources and which is overdosed without an overt, open and transparent rationing mechanism.

4.2. How to Establish a Just Rationing Mechanism?

By what criteria should a public health care system ration its services? As shown in the criticism of egalitarianism, we cannot adopt Daniels' principle (i.e., to guarantee everyone's normal species functioning), or we may overburden society given the scarcity of resources. An alternative is to practice utilitarianism to maximize the utility done to the society, but, as we shall see later, this approach is problematic.

Utilitarian considerations usually loom large in a public health care system. As the resources available for the public health care services are scarce and can not accommodate every demand, it is thus desirable to prioritize the provision of health care services according to their respective medical output. More resources would be devoted to the provision of those resources with a proven record of medical success, whereas those services in the pilot stage would receive less (Ridley, 1998, Ch 10). But it should be noted that a public health care system usually adopts medical utility, not social utility, as the criteria in practising utilitarian rationing. Medical utility is measured by the extent to which a medical objective is satisfied by a certain treatment. Such objectives include the prevention of disease, curing the patient, and controlling incurable diseases and their accompanying pain and suffering. If a particular kind of treatment yields great medical utility, a patient should be treated irrespective of his age, social background and so forth. On the contrary, if a public health care system should maximize social utility instead of medical utility, the elderly, chronic patients and sick prisoners would be disregarded even though they could be effectively served in terms of medical utility. By the same token, those with more to contribute to society would be preferred among others in medical treatment. But this sounds highly unfair (Caplan, 1995, Ch 8).

Admittedly we may continue to uphold the utilitarian principle but modify it by pursuing medical utility rather than the social utility. One way of doing so is to prioritize the provision of services by cost-benefit analysis. The Oregon state government in US adopted this approach in rationing its public health care services and found that curing the common cold is more cost-efficient than the excision of an appendix (Haldron, 1991). This sounds highly counter-intuitive. Many would agree that a number of costly but less efficient treatments deserve low priority. But there are still many treatments (like the transplantation of a liver) which are very costly but regarded as indispensable. So, an alternative approach is to define medical utility in terms of medical effectiveness.

Quality adjusted life year (QALY) is a concept for quantifying medical effectiveness. QALY is attained by devising a quality of well-being' scale (QWB), which measures a patient's syndrome type and level of bodily functioning. Under QWB, "0" means death while "1" represents perfect health. A patient's QALY is calculated by multiplying QWB by the number of surviving years. For instance, when the QWB is 0.5 and the number of surviving years is 2, the QALY is 1. The QALY approach is to prioritize the provision of services in accordance with the increment of QALY.

Nevertheless, the QALY approach is not free from defects. Firstly, is a short but healthier life comparable to a long but less healthy one? Secondly, the QALY approach presumes a QALY is of equal value to anyone. But if A's health is worse than B's, many would prefer A to receive treatment even if the same treatment on B would carry the same QALY. Furthermore, even if B would get more QALY than A would, if the treatment can improve the health of A significantly, people will still prefer A to receive the treatment. This is a blatant deviation from the maximization of the utility principle. Such deviation is not a result of adopting Rawls's Difference Principle because people would prefer giving the treatment to B if A would not benefit significantly from the treatment. There stands another dilemma: if a treatment brings vast utility to a small group of people while another one brings small utility to a vast group of people, which should be chosen? We seem to have no definite criteria to settle the matter, depending on how vast the utility is and how small the group of people is. Different societies view the matter differently, and only a wide enough public consultation could solve the problem (Daniels, 1998b).

Even if the QALY helps in prioritizing the provision of different health care services, it is still unsettled how to demarcate basic and non-basic services. The answer to that depends much on the public concern for health and their willingness in financing the public health care system. It is evident that only extensive public consultation and discussion would settle the problem of what constitutes a basic level of health care services. The Oregon experience revealed that, due to the plurality of values, the public could not reach consensus among themselves regarding the measurement of syndrome categories and the bodily normal functioning. It follows that it is necessary to have extensive public consultation and discussion before a QWB could be formulated for the calculation of QALY.

As suggested above, we do not have definite and objective criteria or a mechanical calculation to prioritize the provision of different health care services. The formulation of QALY requires extensive public consultation and discussion, but eventually QALY serves as one of the important indicators for the prioritization. All along we have to take into account other ethical considerations and only an overt, open and democratic procedure would help in prioritizing the provision of different health care services.

Hence, I recommend the Hong Kong Government to set up a mechanism like the Medical Review Boards (Petrinorich, 1996, Ch 8), which are accountable to the public and should embrace representatives of all walks of life. It should constantly conduct surveys and psychological experiments (Englehart & Rie, 1986) to comprehend the public attitude towards the distribution of health care service so as to prioritize the provision of different services, deriving therefrom a decent minimum of health care. The whole process is to be overt, open and transparent, being receptive to public challenge and providing mechanisms for arbitration. The

Hong Kong Government holds that the primary principle of health care policy is not to allow any people to be denied proper health care service for lack of means. That proper service should be construed as a decent minimum level obtained through the rationing mechanism as long as it is construed in accordance with social justice.

There is a recurring misconception that the medical rationing system's purpose is to curb the growth of the public health care system. For instance, the Oregon rationing programme was conducted under the constraint that public expenditure could increase no more.[5] Although the rationing system would scrutinize the public expenditure on health care services, avoiding expenditures on non-basic health care services and reducing the burden on the public health care system, it does not follow that the public health care system may expand no more. The whole point of the medical rationing system is to set out a publicly endorsed priority of the provision of services. If the public health care system is not financially viable for the provision of those prioritized basic services, it has to resort to the macro-level (the first level) of the distribution of resources so as to strive for more resources for the public health care system. If the Government is not able to accommodate such a demand for greater resources due to the competing interest of other social services like housing, social security and education, we have to, via financing reform measures, increase the government revenue and reduce the expenditure of other public services with a view to enabling the public health care system to provide the prioritized basic health care services.

4.3. Revamping Fees Structure

The public health care system in Hong Kong has recently strengthened its rationing mechanism and has made it more open and accountable, though further improvements are still needed. In the middle of 2002, the Hospital Authority launched a trial in some public hospitals where patients are no longer supplied with non-essential drugs free of charge. Guidelines on the prioritization of services have already been developed by the Clinical Ethics Committee of the Hospital Authority with input from the public. So the health care system has been moving towards the direction recommended in this paper.

However, the problem of sustainability cannot be improved solely by rationing because those who can afford to pay more are still tempted to consume the basic services in the public health care system. As we have seen, over 90% of the hospitalization service is provided by the public health care system, but the user only pays for about 2-3% of the average cost, whereas in Singapore the user has to pay up to 20% of the average cost. The high subsidizing rate in Hong Kong tends to lead to misuse and abuse of services. The values of self-reliance and individual responsibility will therefore be compromised and the public health system will be unsustainable.

In the Health Care Reform Consultation Document published in December 2000, the Government proposed to reform the healthcare system by revamping the fee structure of the public health care sector (Hong Kong Government, 2000). In November 2002, a nominal charge of HK$100 for the accident and emergency service was introduced. In the press release issued by the Hospital Authority on March 29, 2003, it was announced that the fees for in-patient service, specialist out-patient service and general out-patient service would be in effect from April 1. A number of new charges were also introduced, including the hospital admission fee and first attendance fee at specialist out-patient clinics. In the press release issued on April 27, it was announced that a nominal fee of HK$10 will be levied on each drug item as per prescription period.

Raising fees and charges will make those who afford to pay more become more responsible for their own health care. They will either make more contributions to the public health care system by paying higher fees, or reduce the burden by being induced to consume services in the private health care sector as the price gap between the two sectors gets closer. Either way will make the system more sustainable. Although revamping the fee structure can be justified by the values of self-reliance and individual responsibility, it may also pose great financial risks to three vulnerable groups: the low income group, the chronically ill, and elderly patients with little income or assets. As a result, the Confucian ethic of care would be compromised. So the Government has stressed that those on Comprehensive Social Security Assistance (CSSA) would continue to be waived from payment of their medical expenses at public health care services and that there would be a fee waiver mechanism to protect people who are in the three vulnerable groups but are not recipients of CSSA from undue financial burden stemming from their health care needs.

5. CONCLUSION

The public health care system in Hong Kong has been able to meet the basic healthcare needs of the population. So in terms of the Confucian ethic of care, it has been very successful. However, the values of self-reliance and self-responsibility seem to have compromised because the financial accessibility and quality of services of the public health care system are high. This is one of the crucial factors that has made the public health care system become more and more unsustainable. In order to solve the problem of sustainability, the government needs to reform the existing system by strengthening its rationing mechanism and revamping the fee structure in accordance with the Confucian conception of social justice.

City University of Hong Kong
Kowloon, Hong Kong

NOTES

1. The work described in this paper was supported by a SRG (Project No. 700982) and an RCPM grant for the project "An Investigation on the Conception of Social Justice in Taiwan – A Comparative Study" from City University of Hong Kong.
2. This estimation is very controversial and many believe that both the Harvard Report and the Hong Kong Government have overestimated the increase in public health care expenditure (e.g., see Yuen, 1998, 1999).
3. The fieldwork in Beijing and Hong Kong was supported by the SRG grant (Project No. 700982) and the fieldwork in Taipei was supported by an RCPM grant for the project "An Investigation on the Conception of Social Justice in Taiwan – A Comparative Study" from the City University of Hong Kong.
4. Daniels in a later article (1998a) suggested that the patient should be entitled to the health care services that are affordable to the society, further to his claim for normal species functioning. But if the society is not of sufficient resources to provide such services, these two claims would be in conflict.

5. The Oregon medical rationing was, under the restraint that no spectacular growth in the public health care services expenditures would be allowed, to make sure those living below the poverty line could enjoy Medicaid. The purposes of prioritizing the provision of services is to allow the government to draw a line below which no state subsidy would be provided. More people would benefit from Medicaid but the scope of subsidy was narrowed (Hadorn, 1991; Petrinovich, 1996, Ch10). That system still would deny people a decent minimum of health care as some services would be given a lower priority even if they were proven necessary. The poor in Oregon have since had better health care services, but the rationing system did not touch on structural considerations, such as not meeting the ends of social justice (Daniels, 1991).

REFERENCES

Beauchamp, Tom L. & Childress, James F. (1994). *Principles of Biomedical Ethics*, 4th ed. New York: Oxford University Press.

Buchanan, Allen (1981). 'Justice: A philosophical review.' In: Earl E. Shelp (Ed.), *Justice and Health Care* (p. 3-21). Boston: D. Reidel.

Buchanan, Allen (1984). 'The right to a decent minimum of health care,' *Philosophy & Public Affairs*, 13(1), 55-78.

Buchanan, Allen (1995). 'Privatization and just health care,' *Bioethics*, 9(3), 220-239.

Callahan, David (1991). 'Ethics and priority setting in Oregon,' *Health Affairs*, 10(2).

Caplan, Arthur (1992), *If I Were Rich Could I Buy a Pancreas?: And Other Essays on the Ethics of Health Care*. Bloomington and Indianapolis: Indianapolis University Press.

Caplan, Arthur (1995). *Moral Matters: Ethical Issues in Medicine and Life Sciences*. New York: John Wiley & Sons, Inc.

Campbell, Alastair *et al.* (1997) *Medical Ethics*, 2nd ed. Auckland: Oxford University Press.

Confucius (1999). *The Analects*. (Arthur Waley, trans.). Hunan: Hunan People's Publishing House.

Daniels, Norman (1981). Health-care needs and distributive justice,' *Philosophy & Public Affairs*, 10(2), 146-179.

Daniels, Norman (1987). 'Justice and health care.' In: Tom Regan & Donald Van Deveer (Eds.), *Health Care Ethics: An Introduction* (pp. 290-335). Philadelphia: Temple University Press.

Daniels, Norman (1991). 'Is the Oregon rationing plan fair?' *Journal of American Medical Association*, 265(17).

Daniels, Norman (1997). 'Limits to health care: Fair procedures, democratic deliberation, and the legitimacy problem for insurers," *Philosophy and Public Affairs*, 26(4), 303-350.

Daniels, Norman (1998a). 'Is there a right to health care and, if so, what does it encompass?" In: Helga Kuhse & Peter Singer (Eds.), *A Companion to Bioethics* (pp. 316-325). Oxford: Blackwell.

Daniels, Norman (1998b). 'Symposium on the rationing of health care: 2: Rationing medical care - A philosopher's perspective on outcomes and process,' *Economics and Philosophy*, 14(1), 27-49.

Englehardt, Jr., H. Tristram (1981). 'Health care allocations: Responses to the unjust, the unfortunate, and the undesirable.' In: Earl E. Shelp (Ed.), *Justice and Health Care/* Boston: D. Reidel.

Engelhardt, Jr., H. Tristram & Rie, M. A. (1986). 'Intensive care units, scarce resources, and conflicting principles of justice,' *Journal of American Medical Association*, 255, 1159-1164.

Friedman, Marilyn (1987). 'Beyond caring: The de-moralization of gender.' In: M. Haner & K. Nielsen (Eds.), *Canadian Journal of Philosophy*, Supp. Vol. 13, 87-105.

Frohlich, Norman & Oppenheimer, Joe A. (1992). *Choosing Justice: An Experimental Approach to Ethical Theory*. Berkeley: University of California Press.

Garrett, Thomas M. *et al.* (1998). *Health Care Ethics: Principles and Problems*, 3rd ed. New Jersey: Prentice Hall.

Gilligan, Carol (1982). *In a Different Voice*. Cambridge, MA: Harvard University Press.

Graham, Gordon (1988). *Contemporary Social Philosophy*. Oxford: Blackwell.

Hadron, David C. (1991). 'The Oregon priority-setting exercise: Quality of life and public policy,' *Hastings Center Report*, May-June.

Hall, David L. & Roger Ames (1999). *The Democracy of the Dead: Dewey, Confucius, and the Hope for Democracy in China*. Chicago and Lasalle, IL: Open Court.

Hsu, L. (1975). *The Political Philosophy of Confucianism*. London: Curzon Press.

Ho, Lok Sung (1997). *Health Care Delivery and Financing: A Model for Reform*. Hong Kong: City University of Hong Kong Press.

Hong Kong Government (1972). Address by the Governor, Sir Murray MacLehose, at the opening of the 1972-1973 session of the Legislative Council, October, 1972. Hong Kong: Government Printer.

Hong Kong Government (1974). *White Paper on the Further Development of Medical and Health Services in Hong Kong*, Hong Kong: Hong Kong Government Printer.
Hong Kong Government (1993). *Towards Better Health*, Hong Kong: Hong Kong Government Printer.
Hong Kong Government (1994). *Hong Kong 1994*. Hong Kong: Government Printer.
Hong Kong Government (1995a). *Hong Kong: Our Work Together*, Address by Governor Rt. Hon. Christopher Patten at the opening of the 1995-1996 session of the Legislative Council, 11 October, Hong Kong: Government Printer.
Hong Kong Government (1995b). *The Five Year Plan for Social Welfare Development in Hong Kong – Review 1995*, Hong Kong: Government Printer.
Hong Kong Government (1996). *Hong Kong 1996*. Hong Kong: Government Printer.
Hong Kong Government (1997). *Hong Kong 1997*. Hong Kong: Government Printer.
Hong Kong Government (2000). *Lifelong Investment in Health: Concultation Document on Health Care Reform*. Hong Kong: Government Printer.
Hong Kong SAR Government (2001). *Chief Executive's Policy Address*. Hong Kong: Government Printer.
Hospital Authority (1998). *Statistical Report 1996/97*. Hong Kong: Public Affairs Division of Hospital Authority.
King, Ambrose Y. C. (1997). 'Confucianism, modernity, and Asian democracy.' In" Ron Bontekoe and Marietta Stephaniats (Eds.), *Justice and Democracy: Cross Cultural Perspectives* (pp. 163-179). Honolulu, HI: University of Hawai'i Press.
Langan, Mary (Ed.) (1998). *Welfare: Needs, Rights and Risks*. London: Routledge.
Legge, James (trans.) (1960). *The Chinese Classics*. Hong Kong: Hong Kong University Press.
Li, Chengyang (1999). *The Tao Encounters the West*. Albany, NY: State University of New York Press.
Mencius (1999). *Mencius*, (Zhentao Zhao et al, trans.). Hunan: Hunan People's Publishing House.
Miller, David (1976). *Social Justice*. Oxford: Clarendon Press.
Mullen, Penelope (1998). 'Is it necessary to ration health care?' *Public Money and Management*, 18(1), 52-58.
Nielsen, Kai (1989). 'Autonomy, equality, and a just health care system,' *The International Journal of Applied Philosophy*, 4.
Noddings, Nel (1984). *Caring: A Feminine Approach to Ethics and Moral Education*. Berkeley and Los Angeles: University of California Press.
Nozick, Robert (1974). *Anarchy, State and Utopia*. New York: Basic Books.
Okin, Susan (1989). 'Reason and feeling in thinking about justice,' *Ethics*, 99, 39-53.
Petrinovich, Lewis (1996). *Living and Dying Well*. New York and London, Plenum Press.
President's Commission (1983). President's Commission for the Study of Ethical Problems in Medicine and Biomedical and Behavioral Research, *Securing Access to Health Care*, Vol. 1. Washington DC: U.S. Government Printing Office.
Rawls, John (1971). *A Theory of Justice*. Cambridge, MA: Harvard University Press.
Ridley, Aaron (1998). *Beginning Bioethics: A Text with Integrated Readings*. New York: St. Martin's Press.
South China Morning Post (SCMP) (8/12/98). Patients will have to pay in overhaul—Health chief.'
Slote, Michael (1998). 'The justice of ccaring.' In: Ellen Frankel Paul *et al.* (Eds.), *Virtue and Vice* (pp. 171-195). New York: Cambridge University Press.
Social Welfare Department (1999). *Departmental Report 1997/98*. Hong Kong: Government Printer.
Tao, Julia Po-Wah Lai (1998). 'Confucianism.' In: *Encyclopedia of Applied Ethics*, Volume 1. London: Routledge.
Tao, Julia Po-Wah Lai (1999). 'Does it really care?' *The Journal of Medicine and Philosophy*, 24(6),571-590.
Tse, Nancy (1998). 'Counting the cost of health,' *Hong Kong Standard*, 30 March 1998.
Rescher, Nicholas (1966). *Distributive Justice*. Indianapolis, IN: Bobbs-Merrill.
Sunstein, Cass R. (1991). 'Preference and politics,' *Philosophy and Public Affairs*, 20(1), 3-34.
Waltzman, Nancy (1991). 'Socialized medicine now—without wait,' *Washington Monthly*, 23(1).
Walzer, Michael (1983). *Spheres of Justice: A Defense of Pluralism and Equality*. New York: Basic Books.
White, Julie Anne (2000). *Democracy, Justice, and the Welfare State: Reconstructing Public Care*. University Park, PA: The Pennsylvania University Press.
Wilding, Paul, Ahmed Shafigul Hugue, & Julia Tao Lai Po-wah (Eds.) (1997). *Social Policy in Hong Kong*. Cheltenham: Edward Elgar.
World Health Organization (1999). *The World Health Report 1999 – Making a Difference*. Geneva: WHO.

Yuen, Peter P. (1998). 'To change or not to change: A review of health care financing in Hong Kong,"
    Asian Hospital, Third Quarter in Review, 1998, 26-28.
Yuen, Peter P. (1999). Presentation at the Public Forum on Ethics and Policy Choice in Health Care
    Financing, organized by Centre for Comparative Public Management and Social Policy (City
    University of Hong Kong), Hospital Authority of Hong Kong, and Journal of Medicine and
    Philosophy, USA.

NOTES ON CONTRIBUTORS

Alora, Angela Tan, Professor, Southeast Center for Bioethics, Faculty of Medicine & Surgery, University of Santo Thomas, Manila, the Philippines.

Alvarez, Allen Andrew A., Professor, Department of Philosophy, University of the Philippines, Dilliman, the Philippines.

Castro, Leonardo D. de, Professor, Department of Philosophy, University of the Philippines, Dilliman, the Philippines.

Chan, Ho Mun, Associate Professor, Department of Public & Social Management, City University of Hong Kong, Kowloon, Hong Kong.

Dong, Ping, Professor, Department of Social Sciences, Capital Medical University, Beijing, China.

Fan, Ruiping, Assistant Professor, Department of Public & Social Management, City University of Hong Kong, Kowloon, Hong Kong.

Hui, Edwin, Professor of Biomedical Ethics and Spiritual Theology and Dean in charge of the Chinese Studies Program, Regent College, University of British Columbia, Vancouver, Canada.

Ip, Po-Keung, Research Network Associate, Institute of Asian Research, University of British Columbia, Vancouver, Canada.

Kang, Phee-Seng, Associate Professor, Department of Religion & Philosophy, Hong Kong Baptist University, Hong Kong.

Manickavel, V., Professor, College of Medical Science, Kathmandu University, Nepal.

Pak, Un-Jong, Professor of Law, Ewha Women's University, Seoul, Korea.

Qiu, Ren-Zong, Professor, Institute of Philosophy, Chinese Academy of Social Sciences, Beijing, China.

Sakamoto, Hyakudai, Professor, Department of Philosophy, Nihon University, Tokyo, Japan.

Shen, Ming-Xian, Professor, Institute of Philosophy, Shanghai Academy of Social Sciences, Shanghai, China.

Shi, Da-Pu, Professor, Xi'an Medical University, Xi'an, China

Takeuchi, Kazuo, Professor, Kyorin University, 6-20-2 Shinkawa, Mitala-shi, Tokyo 181, Japan.

Tang, Re-Feng, Associate Professor, Department of Philosophy of Science & Technology, Chinese Academy of Social Sciences, Beijing, China.

Tao-Lai Po-Wah, Julia, Director of Centre for Comparative Public Management and Social Policy, Associate Professor, Department of Public and Social Administration, City University of Hong Kong, Kowloon, Hong Kong.

Wang, Xiao-Yan, Professor, Department of Social Sciences, Capital Medical University, Beijing, China.

Wang, Yan-Guang, Associate Professor, Department of Ethics & Center for Applied Ethics, Chinese Academy of Social Sciences, Beijing, China.

Yu, Lin, Editor, *Journal of Medical Ethics*, Xi'an Medical University, Xi'an, China

INDEX

abortion, 3, 14, 15, 18, 25, 39, 49, 50, 53, 54,
  59, 65, 79, 92, 94, 95, 180
advance directives, 19, 59
Africa, 88, 181
AIDS, 8, 79, 192, 197, 199, 200, 201, 202,
  203, 204
  in China, 8
  in the Philippines, 79
aisi (death), 149
alalay (carer), 73
allocation, 13, 16, 20, 57, 77, 221
Alvarez, Allen Andrew A., 6
Alzheimer's, 24
Americans for Immigration Control, 86
anak ni, 73
Analects, 22, 23, 35, 36, 37, 38, 39, 61, 62,
  136, 1391, 142, 143, 149, 215, 216, 217,
  218
animals, 1
  Buddhist attitude towards, 1
  personhood of, 16
arbularyo (herbal medical practitioner), 109
Aristotle, 29, 60
Arthur Kleinman, 6
artificial insemination, 99, 119
Asia(n)
  attitude towards life and death, 97
  bioethics in *See* bioethics, Asian
  concept of personhood, 58
  culture, 2, 5, 41, 57
  ethos, 3, 48, 57
  health care reform in, 8
  morality 57
  values, 5, 101
  view of nature, 47
  way of thinking, 47
ate (older sister), 74
authenticity, 6, 105, 109, 110, 111
autonomia, 17
autonomy, 4, 5, 8, 17, 18, 19, 20, 21, 22, 23,
  24, 25, 26, 41, 48, 54, 59, 86, 95-102, 117,
  175, 198, 201, 203, 204
  Confucian 22

Babinski's reflex, 130
bahala na (come what may), 74, 77, 80
Bangkok, 132
bantay (watcher), 73
Bayles, Michael, 122
Becker, Lawrence, 23
Beijing, 212
Belgium, 132

beneficence, 5, 8, 80, 86, 99, 100, 101, 110,
  179, 191, 193, 198, 201, 202, 204, 214
benevolence, 22, 23, 33, 35, 36, 37, 38, 39, 42,
  50, 53, 54, 136, 137, 139, 143, 144, 145,
  215, 216, 217
bioethics, 2, 3, 4, 5, 6, 7, 13, 18, 23, 25, 33, 40,
  45, 46, 47, 49, 59, 66, 71, 81, 83, 85, 86, 87,
  89, 91, 93, 95, 96, 97, 99, 100, 101, 145,
  146, 190, 196, 198, 200, 204
  American, 59
  Asian, 3, 4, 7, 11, 45, 47, 48, 57, 135
  Chinese, 29
  committee, 131
  communicative, 5, 91, 100-10
  Confucian, 2, 25, 57, 58, 65
  feminist, 25, 101. 102
  Filipino, 4, 71, 72
  global, 1, 4, 7, 47, 48
  in Asia, 2, 3, 5, 7, 45, 57, 66, 91, 135, 144
  local, 4
  principlist, 86, 100, 193
  Reconstructionist Confucian, 4, 66, 67
  rights-based, 4, 57
  Western, 3, 45, 48, 57, 72, 97
Bioethics Advisory Commission, 6
birth control, 25, 174, 174, 181
  policy in China, 174
Book of Documents, 216
Boorse, Christopher, 106
brain death, 7, 92, 129, 130, 131, 132, 133
  in Japan, 129-132
Brain Death and Organ Transplantation
  Committee on Bioethics of the Japanese
  Medical Association, 131
Brazil
  Q-sterilization in, 86
Brihaye, Jan, 132
Buddha, 149
Buddhism, 2, 39, 48, 101, 136, 137
  attitude towards animals, 1
  attitude towards life, 137
  benevolence in 137
  care ethics, 6, 101
  in China, 137
  concept of love, 61
  school, 143
Buenos Aires, Argentina, 45
bukol (lump), 75
Bureau for Maternal and Child Care, Ministry
  of Public Health, 194

Philosophy and Medicine

55. E. Agius and S. Busuttil (eds.): *Germ-Line Intervention and our Responsibilities to Future Generations.* 1998   ISBN 0-7923-4828-1

56. L.B. McCullough: *John Gregory and the Invention of Professional Medical Ethics and the Professional Medical Ethics and the Profession of Medicine.* 1998
     ISBN 0-7923-4917-2

57. L.B. McCullough: *John Gregory's Writing on Medical Ethics and Philosophy of Medicine.* 1998 [CiME-1]   ISBN 0-7923-5000-6

58. H.A.M.J. ten Have and H.-M. Sass (eds.): *Consensus Formation in Healthcare Ethics.* 1998 [ESiP-2]   ISBN 0-7923-4944-X

59. H.A.M.J. ten Have and J.V.M. Welie (eds.): *Ownership of the Human Body.* Philosophical Considerations on the Use of the Human Body and its Parts in Healthcare. 1998 [ESiP-3]   ISBN 0-7923-5150-9

60. M.J. Cherry (ed.): *Persons and Their Bodies.* Rights, Responsibilities, Relationships. 1999   ISBN 0-7923-5701-9

61. R. Fan (ed.): *Confucian Bioethics.* 1999 [APSiB-1]   ISBN 0-7923-5853-8

62. L.M. Kopelman (ed.): *Building Bioethics.* Conversations with Clouser and Friends on Medical Ethics. 1999   ISBN 0-7923-5853-8

63. W.E. Stempsey: *Disease and Diagnosis.* 2000   PB ISBN 0-7923-6322-1

64. H.T. Engelhardt (ed.): *The Philosophy of Medicine.* Framing the Field. 2000
     ISBN 0-7923-6223-3

65. S. Wear, J.J. Bono, G. Logue and A. McEvoy (eds.): *Ethical Issues in Health Care on the Frontiers of the Twenty-First Century.* 2000   ISBN 0-7923-6277-2

66. M. Potts, P.A. Byrne and R.G. Nilges (eds.): *Beyond Brain Death.* The Case Against Brain Based Criteria for Human Death. 2000   ISBN 0-7923-6578-X

67. L.M. Kopelman and K.A. De Ville (eds.): *Physician-Assisted Suicide.* What are the Issues? 2001   ISBN 0-7923-7142-9

68. S.K. Toombs (ed.): *Handbook of Phenomenology and Medicine.* 2001
     ISBN 1-4020-0151-7; Pb 1-4020-0200-9

69. R. ter Meulen, W. Arts and R. Muffels (eds.): *Solidarity in Health and Social Care in Europe.* 2001   ISBN 1-4020-0164-9

70. A. Nordgren: *Responsible Genetics.* The Moral Responsibility of Geneticists for the Consequences of Human Genetics Research. 2001   ISBN 1-4020-0201-7

71. J. Tao Lai Po-wah (ed.): *Cross-Cultural Perspectives on the (Im)Possibility of Global Bioethics.* 2002   ISBN 1-4020-0498-2

72. P. Taboada, K. Fedoryka Cuddeback and P. Donohue-White (eds.): *Person, Society and Value.* Towards a Personalist Concept of Health. 2002
     ISBN 1-4020-0503-2

73. J. Li: *Can Death Be a Harm to the Person Who Dies?* 2002
     ISBN 1-4020-0505-9

Philosophy and Medicine

KLUWER ACADEMIC PUBLISHERS – DORDRECHT / BOSTON / LONDON